Structural Analysis using Computational Chemistry

RIVER PUBLISHERS SERIES IN POLYMER SCIENCE

Series Editors

SAJID ALAVI
Kansas State University
USA

YVES GROHENSE
University of South Brittany
France

SABU THOMAS
Mahatma Gandhi University
India

The "River Publishers Series in Polymer Science" is a series of comprehensive academic and professional books which focus on theory and applications of Polymer Science. Polymer Science, or Macromolecular Science, is a subfield of materials science concerned with polymers, primarily synthetic polymers such as plastics and elastomers. The field of polymer science includes researchers in multiple disciplines including chemistry, physics, and engineering.

Books published in the series include research monographs, edited volumes, handbooks and textbooks. The books provide professionals, researchers, educators, and advanced students in the field with an invaluable insight into the latest research and developments.

Topics covered in the series include, but are by no means restricted to the following:

- Macromolecular Science
- Polymer Chemistry
- Polymer Physics
- Polymer Characterization

For a list of other books in this series, visit www.riverpublishers.com

Structural Analysis using Computational Chemistry

Editor

Norma-Aurea Rangel-Vázquez

PCC, Aguascalientes, Mexico

Published, sold and distributed by:
River Publishers
Alsbjergvej 10
9260 Gistrup
Denmark

River Publishers
Lange Geer 44
2611 PW Delft
The Netherlands

Tel.: +45369953197
www.riverpublishers.com

ISBN: 978-87-93379-95-4 (Hardback)
978-87-93379-96-1 (Ebook)

Contents

Prologue

Computational chemistry is a science that allows to study, characterize, and predict the structure and stability of chemical systems; studying energy differences between different states to explain spectroscopic properties and reaction mechanisms at the atomic level. That is gaining strength day by day due to field application from chemical engineering, electrical engineering, electronics, biomedicine, biology, materials science, to name a few. This book arises from the need to present the progress of computational chemistry in various application areas.

In Chapter 1, using hybrid molecular mechanics, structural properties for the synthesis of a hydrogel that can be used as part of a transdermal patch chitosan cross-linked with genipin for controlled release of glibenclamide were obtained and currently has several drawbacks, as are poor control of drug in the plasma or patient discomfort. Hypoglycemic (glibenclamide) is used for the treatment of diabetes type II in oral form, which already represents in Mexico as one of the three leading causes of death.

In Chapter 2, the adsorption of metformin study, drug used in conjunction with glyburide for the treatment of diabetes type II, was verified by Gibbs free energy negative which was verified by identifying nucleophilic areas and electrophilic of the MESP as well as molecular vibrations that originated in the FTIR.

In Chapter 3, due to in Mexico, about 400,000 children under 15 years with diabetes type I, insulin properties in order to develop a controlled release patch through chitosan were studied. molecular mechanics model employment by AMBER because insulin is a polypeptide of 51 amino acids, thereby obtaining the structural properties of the adsorption process.

In Chapter 4, because solar cells and especially organic solar cells represent a way of obtaining energy that will be required in the future, the need to obtain new materials more resistant to the degrading environmental agents to allow the durability of OPV and these can be incorporated into different devices for everyday use. Therefore, computational chemistry allowed for the structural analysis of nanocomposites-based fullerene to be used in the design of new solar cells.

In Chapter 5, using modeling Monte Carlo computational chemistry, thermodynamic properties of ionic liquids were calculated using the molecular description of all the atoms in order to obtain very small margin of deviation compared with the experimental values of the thermodynamic properties of a system, so that prediction methods and computer simulation (molecular simulation) are ideal for estimating material properties design tool, thereby avoiding the great difficulty of synthesis in the laboratory and the cost for derivatization.

Acknowledgments

To God, for giving me the opportunity to live and learn every day from my mistakes, to be a better person.

My father, Ricardo Rangel (+), for guiding me in every moment of life anywhere you are.

My family, to dream with me and teach me that with hard work and dedication can achieve the goals very even though the adversities of life, how hard is worth.

My love, The support of the person you love is a great blessing . . . Thank you for being part of that force that drives me going, because no matter the distance that separates us, there will always be a sky that unites us, because I was born to adore you, love you and I vowed not to leave you and I will never leave you . . . LOVE YOU, ALWAYS, HEART.

My students, Nancy, Claudia, Ana Karen, and Sofia, for the dedication and effort in research projects that led to the publication of the same, as well as friendship during this time.

My friend Elena Smith and her family (Isaac and Gabby), thank you for your friendship and advices in every moment in the last years.

Thank you

List of Contributors

Ana-Karen Frías-González, *Chemical Engineering Department, Master in Science in Chemical Eng. Technological Institute of Aguascalientes, Mexico*

Claudia-Lizeth Salas-Aguilar, *Chemical Engineering Department, PhD in Engineering Sciences, Technological Institute of Aguascalientes, Mexico*

Ediht-Sofia Martínez-Rodríguez, *Chemical Engineering Department, Master in Science in Chemical Eng. Technological Institute of Aguascalientes, Mexico*

Nancy-Liliana Delgadillo-Armendariz, *Chemical Engineering Department, PhD in Engineering Sciences, Technological Institute of Aguascalientes, Mexico*

Norma-Aurea Rangel-Vázquez, *PCC, Aguascalientes, Mexico*

List of Figures

List of Tables

List of Abbreviations

A	Helmholtz free energy
Å	Angstrom
AA	All atoms
AMBER	Assisted Model Building with Energy Refinement
AM1	Austin Model 1
C	Chitosan
Cm	Centimeters
D	Dielectric constant
ΔHf	Enthalpy of formation
DFT	Density functional theory
E	Total energy
ε	Minimum energy
eV	Electronvolts
FTIR	Fourier transform infrared radiation
G	Degree of cross-linking
Ge	Genipin
$G°f$	Gibbs free energy
H	Hamiltonian operator
HDPE	High-density polyethylene
HOMO	Highest occupied molecular orbital
i	Imaginary unit
Kcal	Kilocalories
*k*e	Stretching force constant
LJ	Lennard-Jones
Log P	Logarithm of the partition coefficient
Log S	Logarithm of solubility coefficient
LUMO	Lowest unoccupied molecular orbital
m	Mass
MC	Monte Carlo
MD	Molecular dynamics

MESP	Electrostatic Potential Map
MINDO	Modified Intermediate Neglect of Differential Overlap
MM	Molecular mechanics
MP2	Moller–Plesset perturbation theory: 2
MS	Semiempirical methods
MW	Molecular weight
NDDO	Neglect of diatomic differential overlap
Qi and Qj	Atomic partial loads
PCPDTBT	Poli[4,4-bis(2-etilhexil)ciclopenta[2,1-b;3,4-b_]ditiofeno-2,6-diyl-alt-2,1,3-benzotiadizol-4, 7-diyl]
PM3	Parametric method 3
PM6	Parametric method 6
Q/Ge-G	Chitosan/Genipin-Glibenclamide
QM	Quantum mechanics
QSAR	Quantitative Structure–Activity Relationship Properties
r	Distance between the two atoms
$r$eq	Equilibrium length,
$r$ij	Distance separating the two nuclei in a
r0	Sum of the VDW radii
T	Kinetic energy potential energy
U	Potential energy
UA	United atoms
v	Speed
VDW	Van der Waals
ψ	Wavefunction
∇^2	Laplacian operator

1

Quantum Mechanics and Structural Molecular Study (AM1)

Norma-Aurea Rangel-Vázquez[1]
and Nancy-Liliana Delgadillo-Armendariz[2]

[1]PCC, Aguascalientes, Mexico
[2]Chemical Engineering Department, PhD in Engineering Sciences, Technological Institute of Aguascalientes, Mexico

Abstract

Chitosan has been characterized as a polymer with excellent characteristics for use in the medical area due to compatibility with living organisms, and it is harmless, easy to acquire, and biodegradable. It comes mainly of chitin, which in turn is extracted from crustacean shells. Use as releasing hydrogels forming drug is common; however, to obtain a stable hydrogel is necessary to use cross-linkers, which are substances that form a kind of network that keeps firm and in balance. Among the most common cross-linkers are aldehydes, although the genipin may soon be number one because it is a natural compound and appropriately intertwined with chitosan. These hydrogels are used in transdermal patches for the application of drugs in a controlled manner, which have been used a variety of drugs; however, they may still include many which are used in large quantities and have the appropriate characteristics, such as hypoglycemic (glyburide), which is used to treat type II diabetes mellitus, currently considered a global epidemic.

Furthermore, computational chemistry allows among other things known properties of pure substances or joined as in the case of this type of hydrogel, obtained by the analysis of the information relevant to the synthesis chemical structure. There are several models such as the so-called semiempirical methods, which yield good results with little computational expense.

Keywords: AM1, chitosan, glibenclamide, genipin.

1.1 Theoretical Basis of Quantum Mechanics

It is necessary to establish that the two main ways to obtain molecular properties using computational chemistry are molecular mechanics (MM) and quantum mechanics (QM). At first, the molecules are analyzed as a formation of spheres and springs, which are recognized as springs and mathematical algorithms follow Hooke's law and use the force fields, where there are many parameters and approximations.

In contrast, QM is based on subatomic particles, explicitly treating the electrons, which have a nature of wave–particle dualism. The QM studied molecules as sets of atoms with their specific electronic envelope, bases its calculations on the energy states of the interactions between orbitals, and describes both the bonds that are formed as those (chemical reactions) are broken, systems typically small, and have a high consumption of computing resources.

The QM considered a molecule as a system where there are atomic orbitals, which in turn originate molecular orbitals. These molecular orbitals can have different shapes and orientations; they are an energy space where there is greater likelihood that a specific electron is present and represents the area of greatest interest. To describe this system, the QM believes that the cores have almost no movement and form a field called *core*, where subatomic particles exempting electrons, which form paths around this. Thus, one application of chemical or QM is the study of the electronic behavior of atoms and molecules, and their relationship to the molecular structure and chemical reactivity.

Furthermore, according to the hypothesis that electrons have wave properties, it may be describable from mathematical equation waves, particularly of the standing wave, thus establishing the call wavefunction $[\psi(x, t)]$, which describes the evolution of motion of the electron around the core [1].

In the QM, system is completely determined by the solution of the corresponding wavefunction in the same way that a thermodynamic system is defined with the knowledge of three state variables. A wavefunction represents an entity or unit of a Hilbert space of infinite dimension that includes all possible states of the system. However, it was the Austrian physicist Erwin Schrödinger who applying this wavefunction a series of mathematical operations (Hamiltonian operator) established the basis of the QM expression.

The Schrödinger equation or equation wavefunction is a vector formed by partial differential equation and linear operators giving a general solution by adding specific solutions. For this equation, it is considered a given volume of the system under study. This system has the characteristics that describe called system status. A state of n particles comes completely described by the wavefunction (see Equation 1.1):

$$\psi(\vec{r_1}, \ldots, \vec{r_n}; t) \tag{1.1}$$

This wavefunction provides the time evolution of the state of a system. Physical information contained therein is known through its modulus squared (Equation 1.2):

$$|\psi(\vec{r}, t)|^2 \;{}^\wedge 2 \tag{1.2}$$

The squared modulus of the wavefunction (Equation 1.3) is the probability per unit volume of finding a particle. This amount is called probability density.

$$P(\vec{r}, t) = |\psi(\vec{r}, t)|^2 \tag{1.3}$$

Since the wavefunction is complex, usually this expression means (Equation 1.4):

$$|\psi(\vec{r}, t)|^2 = \psi * \psi \;{}^\wedge 2 \tag{1.4}$$

If desired, locating the particle in a volume *V*, the Equation 1.5 is effected:

$$\int_{-\infty}^{\infty} \psi^* \psi \cdot dV = 1 \tag{1.5}$$

This condition is essential for a function that can be considered representative of a quantum state in a system, for example, an electron. The main purpose is then to describe the state of a system, being obtained by applying the wavefunctions $\psi(\vec{r}, t)$ either stationary or over time. Furthermore, the magnitude which can predict the behavior of the system is the total energy, which remains constant over time, but can be transformed into different types. This is contained in the Hamiltonian applied to the wavefunction, which is represented as follows:

a) Equation 1.6 represents a moving particle entailing a kinetic energy:

$$T = \frac{mv^2}{2} \tag{1.6}$$

where *m* is the mass and *v* is the speed.

b) A quantity called linear momentum of the particle, which is appreciated in Equation 1.7, has

$$p = m\ v \tag{1.7}$$

Therefore, kinetic energy is expressed as follows in Equation 1.8:

$$T = \frac{p^2}{2m} \tag{1.8}$$

c) In the case where the system has an electron moving around a nucleus, this presents a potential energy (Equation 1.9):

$$U = U(\vec{r}) \tag{1.9}$$

It can be seen that this energy depends only on the position of the particles involved. In Equation 1.10, electrical potential energy between protons and electrons is observed:

$$U = K\frac{q_p q_c}{r} \tag{1.10}$$

Thus, in Equation 1.11, we have the sum of kinetic and potential energy to be constant in time, which is called total energy system:

$$E = T + U \tag{1.11}$$

Both kinetic energy T and potential energy U (total energy E) are observable, i.e., they are measurable in the systems; therefore, they can be associated quantum operators (Equation 1.12):

$$H = T + U \tag{1.12}$$

Because the temporal evolution of a system must be such that the total energy remains constant, the total energy equation (Equation 1.13) condenses all the information on the possible evolution.

$$E = T + U(= H) \tag{1.13}$$

Thus, the Schrödinger equation (Equation 1.14), representing the wavefunction and the energy of a molecule, is represented as follows:

$$\widehat{\mathrm{H}}\psi = \mathrm{E} \tag{1.14}$$

where "ψ" is the wavefunction that describes the state of the system and all its properties, and $\widehat{\mathrm{H}}$ is the Hamiltonian operator and consists of a set of activities within several terms where the kinetic energy and potential energy are represented operations.

Finally, E symbolizes the energy associated with the Hamiltonian and is quantized energy system and its own value. Therefore, the above equation can be written as follows (Equation 1.15):

$$-\frac{h^2}{2m}\nabla^2\psi(\vec{r},t) + U(\vec{r},t)\psi(\vec{r},t) = ik\frac{\partial}{\partial t}\psi(\vec{r},t) \tag{1.15}$$

where

m = the mass of the particle,
∇^2 = Laplacian operator,
i = imaginary unit, and
$\hbar$ = normalized Planck constant ($h/2\pi$).

It is noteworthy that the Laplacian operator represents the sum of partial derivatives shown in Equation 1.16:

$$\frac{\partial^2}{\partial x^2} + \frac{\partial^2}{\partial y^2} + \frac{\partial^2}{\partial z^2} \tag{1.16}$$

The first term of the Schrödinger equation represents the kinetic energy, while the second term represents the potential of moving energy molecule, forming the Hamiltonian which is applied to the original wavefunction. In this equation, it is considered a potential energy dependent on time. Equation 1.17 shows the independence of time for a dimension:

$$-\frac{h^2}{2m}\frac{\partial^2\psi(x)}{\partial x^2} + V(x)\psi(x) = E\psi(x) \tag{1.17}$$

where V equals the potential energy with respect to "x" applied to the wavefunction. Similarly, the Hamiltonian can be represented by the Equation 1.18:

$$\hat{H} = -\underbrace{\frac{\eta^2}{2}\sum_a \frac{1}{ma}\nabla_a^2}_{1} - \underbrace{\frac{\eta^2}{2m}\sum_i \nabla_i^2}_{2} + \underbrace{\sum_a\sum_{\beta>a}\frac{Z_a Z_\beta e^2}{r_{a\beta}}}_{3} - \underbrace{\sum_a\sum_i\frac{Z_a e^2}{r_{ia}}}_{4} + \underbrace{\sum_i\sum_{i>j}\frac{Z_{e^2}}{r_{ij}}}_{5} \tag{1.18}$$

where terms (1) represents the kinetic energy of the nuclei, (2) represents the kinetic energy of electrons, (3) represents the repulsion of the nuclei,

(4) represents the attraction core-electron, and finally (5) represents electron repulsion, corresponding:

α, β = nucleus,
i, j = electrons,
z = atomic number, and
r = distance between particles.

However, the complexity of the molecules makes it practically impossible to solve the equation of wavefunction, this being possible only through approximations. The valence bond theory, functional theory, and molecular orbital are some of the tools for their solution [1, 2].

1.1.1 Semiempirical Methods

The solution to the equation wavefunction is achieved through various models called AB-initio (from the beginning) to obtain the most accurate results. However, due to the large computational cost and considering the objectives of each project and the type of molecules or systems being studied, among other criteria, you can opt for the use of models called semiempirical, which are usually found part of computer suites such as Gaussian, Spartan, or HyperChem, among others.

These models base their calculations on the valence electrons and although its foundation comes from AB-initio, to get the results in a more agile way to calculate several comprehensive omitted and instead take into account values or parameters achieved so and annex experimental method. Through these, models can be obtained properties such as total shit, molecular energy, atomic electron density, polarization, or vibrational spectra, to mention a few. Among the semiempirical methods, there is a very popular group from work Dewar: the MINDO, AM1, and PM3. There is currently a model called PM6 that already contains visible corrections to obtain better results for specific molecules compared to previous methods, such as heat of formation.

1.1.1.1 The semiempirical method AM1

This model, named in honor of the University of Texas at Austin (Austin Model 1), corrects and reparematrizes the theoretical mathematical logarithm which is the basis in the MNDO method without increasing the calculation time. Add the Gaussian functions to improve results where hydrogen bonds involved and at the same time increasing the number of parameterized data. AM1 parameterized for H, B, Al, C, Si, Ge, Sn, N, P, O, S, F, Cl, Br, I, Zn, and Hg. This method replaces the terms of nuclear repulsion on the potential energy to parameterize terms of core repulsion in the Schrödinger equation. The terms compensate for the fact only consider the valence electrons in the equation, and they incorporate electron correlation effects. Thus, according to Dewar, the model AM1 represents great importance for organic compounds, has as advantages a better reproducibility of the hydrogen bonds, and has a better estimate of the activation energy [3].

1.1.1.1.1 *Application of AM1 method in molecular structural study*

Semiempirical methods in QM have great advantages over AB-initio methods; the main features are the notable reduction in computation time and accuracy of the results, for which it is essential to select exhaustively a good model for calculation together with the base and suitable for each system parameterization.

Next, a structural analysis is shown by AM1 model of individual molecules and connected to the main components of a polymer matrix to release the drug glibenclamide and the water molecule to be absorbed by the matrix which can make it hydrogel. The distinctive colors of the elements used are as follows:

● oxygen, ● carbon, ● nitrogen, ○ hydrogen, ○ sulfur, and ○ chlorine.

1.1.1.1.2 *Certain molecular properties*

a) The Quantitative Structure–Activity Relationship (QSAR) Properties

It is a mathematical hypothesis based on the fact that the structure of a molecule is primarily responsible for its chemical, physicochemical, biological, or pharmacological properties [4, 5]. Perhaps one of the fundamental premises of the theory is the principle of structural similarity, which states that similar molecular structures have similar properties, while different molecular structures demonstrate different properties. While it is known for a long time that different substances have different biological effects, progress in determining structures allowed establishing structure–activity relationships (SAR), which show certain effects on the biological activities from the change in the chemical structure of a particular compound.

Thus, the QSAR theory seeks to quantify the SAR relations through the development of models and methods combined with mathematical statistics of computational chemistry [6]. They provide data as the molecular surface area, volume, or the log P (logarithm of the partition coefficient), whose values can be positive or negative indicating a lipophilic or hydrophilic character of the molecule, respectively.

b) Molecular Energy

Determining the energy in the molecules is extremely important because it indicates among other things, behavior, affinity, and molecular evolution of the system.

The molecules have different types of energy and depend on the goals, and approaches of each type of energy research are necessary to determine. The energies are calculated: the internal energy (U) indicating the total energy existing in the system at the molecular level, the Helmholtz free energy (A), which identifies the spontaneity of chemical reactions indicating the affinity between the elements, the binding energy (E) which is proportional to the stability of the molecular system, and the enthalpy of formation (ΔHf) showing endothermic or exothermic chemical reaction degree or given binding.

c) Infrared (IR) Spectrum

A properly optimized structure, with distribution of calculated loads, can serve as a starting point for a vibrational analysis of normal coordinates, that is, to predict the vibrational spectra (IR) [7]. It can be carried out using the structure, energy, and force constants. In the semiempirical methods, force constants are taken from a bank of experimental data in AB-initio methods which are calculated from the second derivatives of energy with respect to the coordinates of vibration [8]. The analysis of infrared radiation (IR) is a method dealing with specific spectroscopic infrared part of the electromagnetic spectrum. It is used to identify a compound and investigate the composition of a sample.

The Fourier transform infrared radiation (FTIR) uses infrared radiation to record molecular movements through computer programs. Formula called Fournier transform and conversion scheme called Michelson interferometer are used. To measure a sample, a beam of infrared light passes through this, and when the excitation frequency of a bond (or group of bonds) matches any of the frequencies included in the component waves, ray absorption occurs. What is to be registered is the amount of energy absorbed at each wavelength. This can be achieved by scanning the spectrum with monochromatic beam, which changes wavelength over time or using a Fourier transform to measure all wavelengths at once. From this, one can trace a transmittance or absorbance spectrum, which shows in which wavelengths the sample absorbs IR and allows an interpretation of what bonds are present [9].

d) Electrostatic Potential Map (MESP)

It is a map constructed from different loads present in the molecule or system indicating the existing electron density. The areas of highest electron density is often shown in red, while the free zone electrons are shown in blue and represents the area most likely to nucleophilic attack, or what is the same has an affinity for electrophilic elements.

1.1.2 Computational Suite (HyperChem)

HyperChem is a computational molecular modeling package with tools for chemical simulation and calculation and visualization of molecular properties. Methods of QM, MM, and molecular dynamics are complemented by 3D visualization and animation. Among the methods of QM with which account are the AB-initio and semi-empirical methods such as AM1 or PM3 [10]. This software package is used to determine the molecular properties with the selected model AM1.

1.1.2.1 The molecules analyzed

- Chitosan: Natural Polymer

Polymers are macromolecules that are obtained by repetitions of one or more single units called monomers bonded together by covalent bond [11]. Natural polymers are polymers or biopolymers of natural origin that are characterized by complex structures, low polydispersity, and easy process of degradation [12].

Chitosan is characterized by being biocompatible [13] and is often used to cause drug-eluting hydrogels, which can be physically or chemically intertwined for better quality in their mechanical properties. Chitosan is derived from chitin, which go to deacetylating to form the composite polymer known as chitosan.

The final degree of acetylation and the location of these functional groups are the most important parameters and determine the functional and physiological properties of chitosan as is its degree of solubility. Chitosan consists of the following elements: carbon, nitrogen, oxygen, and hydrogen. It is noteworthy that as shown in Figure 1.1, the acetyl group is positioned on the first ring of chitosan, and the experimental analysis has observed the influence of this functional group depending on its location in the molecule. Figure 1.2 shows the chemical structure of chitosan represented by a unit acetylated and deacetylated. The percentage of deacetylation was established as 75%.

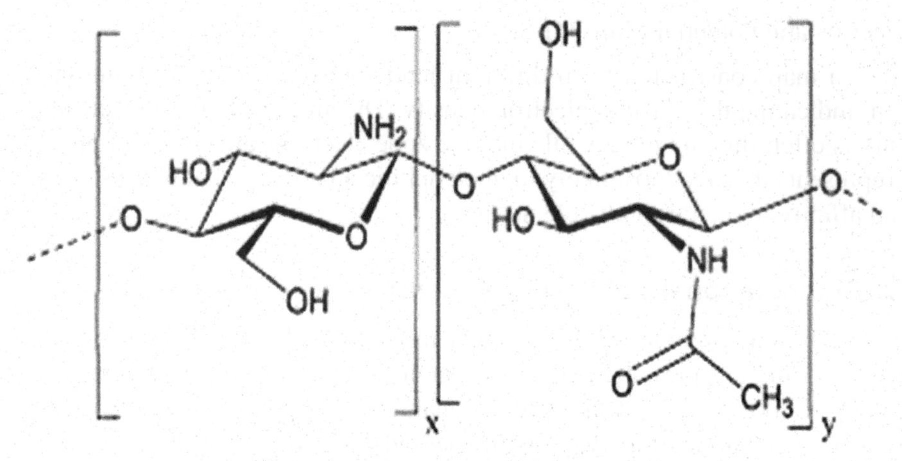

Figure 1.1 Chemical structure of chitosan.

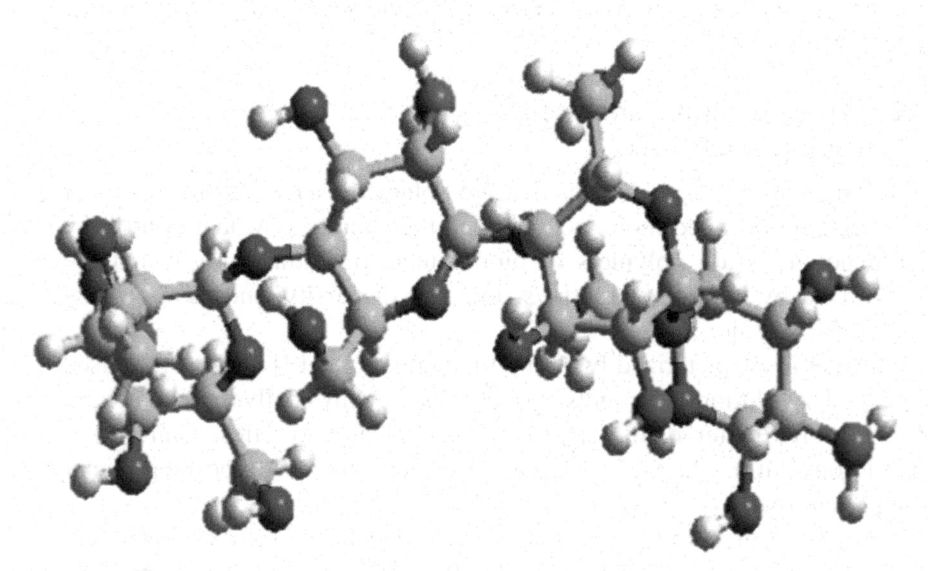

Figure 1.2 Chemical structure of chitosan by HyperChem software.

- Genipin: natural cross-linking

A cross-linker is a substance that forms a three-dimensional network in the hydrogel or polymer matrix formed which maintains stable and balanced structure.

These can be synthetic or natural as glutaraldehyde as genipin obtained from *gardenia flower*. It is noteworthy that the degree of cross-linking is important to achieve a good swelling of the hydrogel and a correct drug release. This cross-linker is highly chitosan used as the cross-linking much favored. The amino group of chitosan binds with genipin methyl group, forming a new amide group. In Figure 1.3, it can be seen that its chemical structure contains a methyl ester group attached to a diene formed by two heterocycles, and $C_{11}H_{14}O_5$ is the chemical formula, and in Figure 1.4, the molecule drawn by HyperChem is shown.

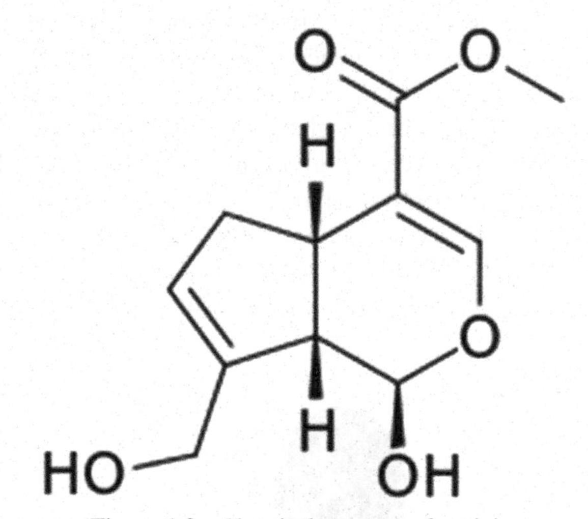

Figure 1.3 Chemical structure of genipin.

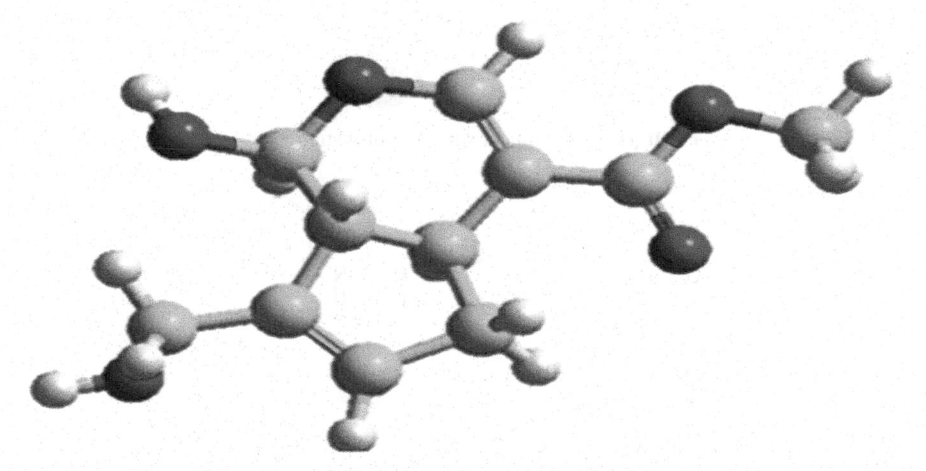

Figure 1.4 Chemical structure of genipin by HyperChem software.

- Water

The water molecule is small, just 18 amu; however, it has fascinating qualities. Water is the main candidate to join as polar solvent in a polymer matrix, thereby making it a hydrogel, in addition to being the main route of drug transport; that is why its analysis is also included. In Figure 1.5, unoptimized structure and optimized AM1 areshown.

- Glibenclamide

Glibenclamide is a widely used drug in the treatment of type II diabetes mellitus, which is a multifactorial, degenerative disorder that is characterized by mismanagement that exists on the amount of sugar in the blood, keeping abnormally high levels and causing a chronic deterioration, that is why it is necessary to use as hypoglycemic glyburide, which lowers the levels of blood sugar when diet or exercise has failed. As shown in Figure 1.6, it is mainly composed of benzene rings, cyclohexane, sulfonyl group, and amines. It has a chemical formula of $C_{23}H_{28}N_3ClO_5S$ and a molecular weight of 494,004 g/mol. In Figure 1.7, the molecule drawn by HyperChem is shown.

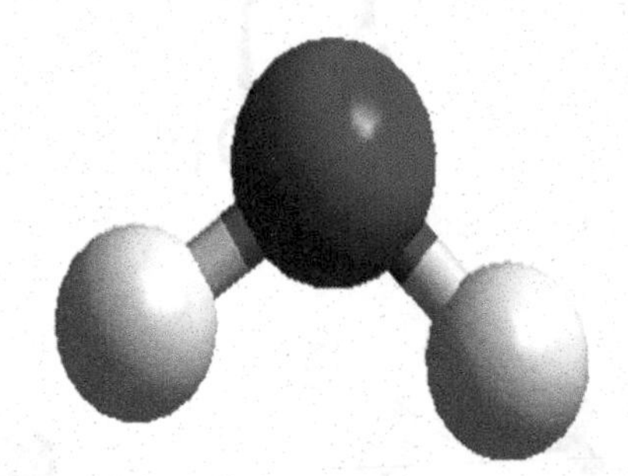

Figure 1.5 Chemical structure of water.

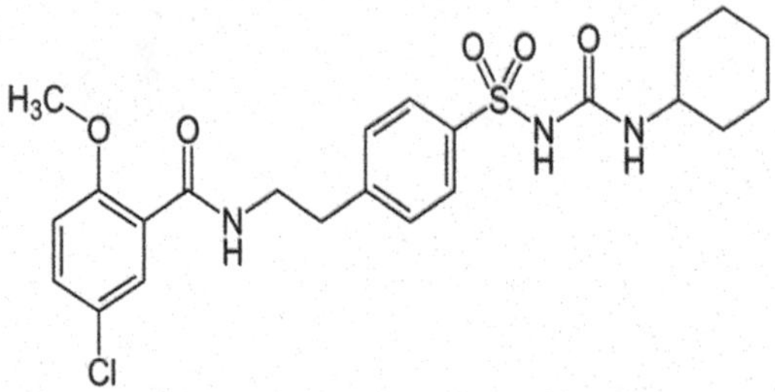

Figure 1.6 Chemical structure of glibenclamide.

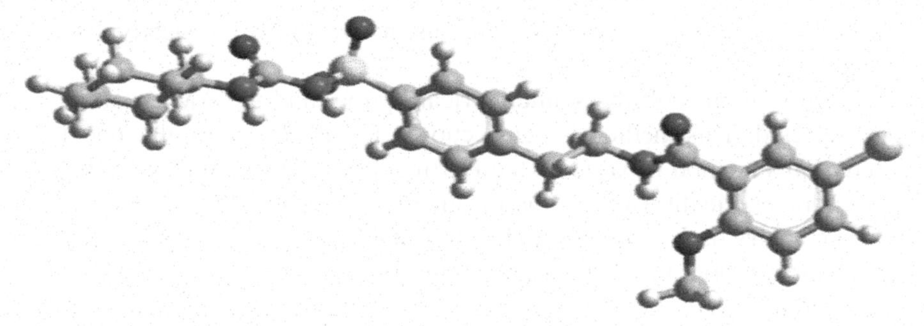

Figure 1.7 Chemical structure of glibenclamide by HyperChem software.

1.1.2.2 Molecular modeling

To achieve the molecular models presented, the following methodology is applied:

- By professional V.7.0 HyperChem software package with semiempirical model option, AM1 method begins by selecting from the menu bar to determine the structural properties of each of the molecules under study; first, the outline of the molecule is made, which is achieved by selecting one by one the elements that constitute the draw tool, located in the toolbar (tool icons), and based on the structure of Lewis, considering the type of atoms and bonds present.
- For the construction of the double bonds in the rings located, double click on the appropriate link for the case where the existence of a circle resonance occurs inside the ring it is given.
- To ensure the proper construction of the molecule is found that the atomic mass calculated by the program is the same as that reported in the literature. The molecules can be represented in different designs; however, first select the lines of sticks and then the balls, rods, and cylinders to better appreciate the connection types, the resonance rings, and complementary hydrogens [14].

1.2 Calculation of Molecular Properties

1.2.1 Molecular Energy

The values of molecular energies on study (glibenclamide, chitosan, genipin, and water) are calculated as follows:

- Using the display tool and the rendering command with the balls and cylinders option are selected. With this option, the molecule takes the form of a ball and stick structure with distinctive colors for each type of atom.

- Using the setup tool, the semiempirical option, and the AM1 method are selected. This method is being chosen for all the molecules under study. This method is based on the Hartree–Fock algorithm to solve the Schrödinger equation and to find the energy and system locations.
- The geometry of the molecule to be more stable by the compute tool and the geometry optimization option is optimized.
- Once optimized the geometry energies, associated molecule is analyzed. The binding energy appears at the bottom of the status line screen; however, you can read each of the energies in the compute/properties option, where the total energy of the system is selected and the various energies of the molecule are read in units of Kcal/mol. These energies are binding energy (E), heat of formation (ΔH), Helmholtz free energy (A), and total energy (U).

1.2.2 Obtaining the QSAR Properties

The purpose of calculating the QSAR (quantitative structure–activity relationship) properties is quantitatively correlate structural features with the physicochemical properties of the molecules under study and get to know, among other features properties, whether these can be related to a polar solvent, determine the area and volume of the molecule and pharmacokinetic properties, mainly. Thus, it is used in the menu compute/QSAR properties with which the properties of the surface area, volume, mass, and log P (log partition coefficient) are obtained from each of the molecules, where the latter property indicates the degree of hydrophilicity or hydrophobicity of a substance.

1.2.3 FTIR Analysis

For the FTIR spectra of individual molecules and determine the wavelengths option Compute/vibration rotation analysis is selected, in a range of 6000–200 cm^{-1} and then the vibrational spectrum option with which the spectrum is analyzed FTIR in various vibrations selecting a certain frequency. By selecting the signal on the spectrum, it shows the bond or bonds that correspond to the vibrational mode.

1.2.4 Electrostatic Potential Map

The map of the electrostatic potential of the various structures is calculated with the option 3D rendering Mapped Isosurface located in the menu bar inside the tool compute, which provides a graphical 3D contour and indicates the site of higher or lower electronic density by color differences, which can be determined by more susceptible electrophilic (red) and nucleophilic (blue) attacks areas. It is noteworthy that these analyses allow us to observe how molecules attract each other, whereas the part of the molecule with excess electrons will join with the poor area of these, or in other words, areas that could attract positively charged particle (proton) will be assigned areas such as higher electron density and vice versa. The more intense the color is, the more nucleophilic or electrophilic property the area has, when a neutral zone is presented in green.

1.2.5 Determination of Glibenclamide/Water Solubility

To calculate the affinity of glibenclamide by a polar solvent such as water, and the appropriate concentration, QSAR values calculated in the molecule glibenclamide, mainly the partition coefficient (Log P) indicating they used the degree of hydrophilicity of drug and shows affinity to form a solution. With this value, the solubility coefficient (Log S) to estimate the saturation/concentration of a chemical in water is determined. This property is often calculated using the octanol/water partition using Equation 1.19 [14]:

$$\text{Log}\, S = 0.796 - 0.854 \log P - 0.00728\ (\text{MW}) + \Sigma\, hj \tag{1.19}$$

where

S = water solubility (mol/L),
P = partition coefficient (octanol/water),
MW = molecular weight of the substance, and
hj = correction factor for each functional group.

Later, with the molecular mass of glibenclamide, suggested grams of the drug are determined in a liter of water, to finally analyze the energy of this union, the electrostatic map, and FTIR spectrum, which they are calculated according to the procedure described above.

1.2.6 Degree of Cross-Linking in the Polymer Matrix

To perform the test is considered a cross-linking covalent type with an acidic solvent and a degree of cross-linking of 33%, this being a common percentage in the synthesis of hydrogels by the good results in the swelling and drug release, where during hydrogel formation transformation chitosan amino group whereas an acidic solvent is the reaction initiator and proton donor is analyzed. This degree of cross-linking is determined by Equation 1.20.

$$G(\%) = \frac{\text{inicial} - \text{final}}{\text{inicial}} \times 100 \tag{1.20}$$

where

G (%) = degree of cross-linking,
NH_2 Inicial = number of free amino groups in the sample uncross-linked, and
NH_2 final = number of free amino groups in the sample cross-linked.

It is noteworthy that the higher the degree of cross-linking, the lower rate of drug is released due to increased density of the matrix [15].

1.2.7 Covalent Cross-Linking

For the analysis of cross-linking of chitosan with genipin (Figure 1.8), first covalent bonding (chemical reaction) which takes place between these two compounds is determined, which starts by a proton donor solvent acid, such as acetic acid. The amino group of chitosan binds with the genipin methyl forming an amide group, and as the percentage of cross-linking of chitosan is 33%, it is considered a molecule of genipin and two chains of four repeating units of chitosan.

1.2.8 Polymer Matrix/Glibenclamide

To analyze the properties of the matrix with glibenclamide, energy calculations, QSAR, FTIR properties, and electrostatic potential are performed. In Figure 1.9, the cross-linked matrix with genipin and glibenclamide molecule is observed. They are drawn with a random orientation in 3D and with a presentation of balls and rods.

1.3 Results

1.3.1 Structural Analysis of Glibenclamide (G)

1.3.1.1 QSAR properties and energy

Table 1.1 shows the data obtained through the QSAR properties observed. The dimensions of the molecule and Log P, which is 3.5, indicating that it is a lipophilic molecule, are observed. However, for the value it presents, this does not mean it is completely insoluble in water, but rather that is poorly

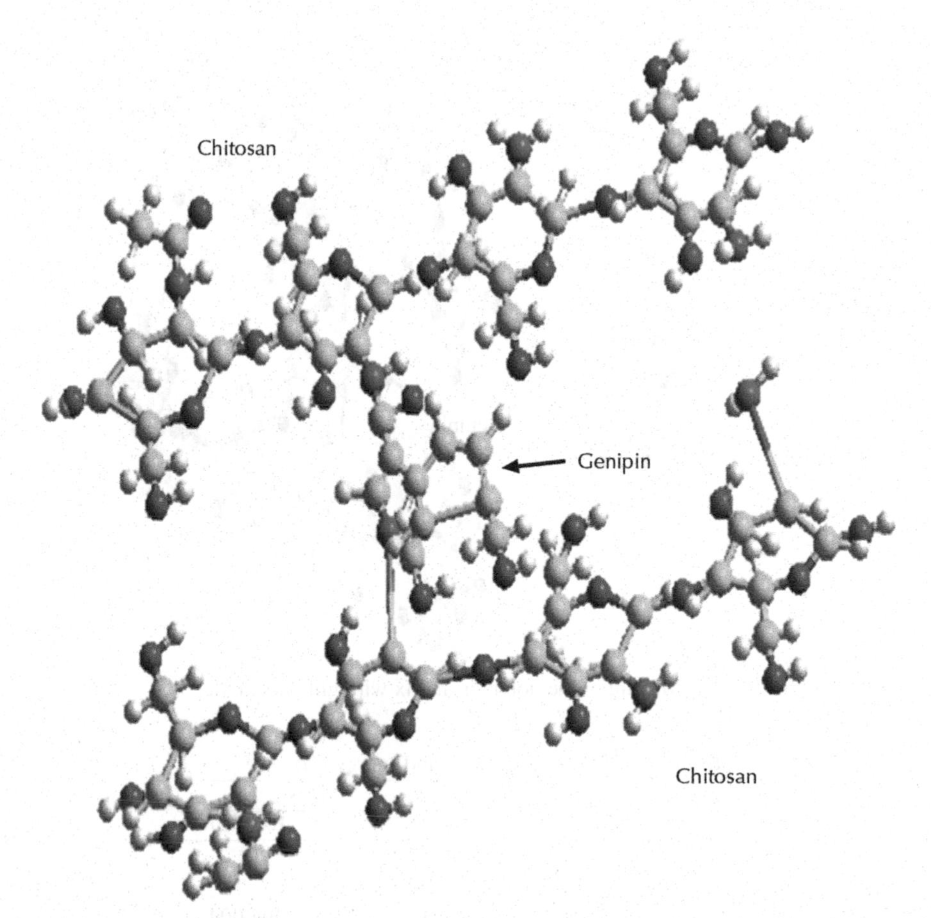

Figure 1.8 Chitosan–genipin cross-linking (HyperChem).

soluble in it, indicating that a large amount of solvent and small amount of drug in the solution should be used, which will be valued the use. The data obtained by AM1 are similar to those provided by other software packages and reported experimental data, where, for example, the Log P was calculated with values of 4.7 and 3.7 according to Shuster et al. and Marvin software, respectively [16].

Moreover, the value of $E = -6277.66$ Kcal/mol indicates a binding energy between the bonds composing, and heat of formation (ΔHf) shows an exothermic negative origin during the formation of the molecule. Also, the free energy (A) having is negative, whereby the spontaneity of the reaction is checked.

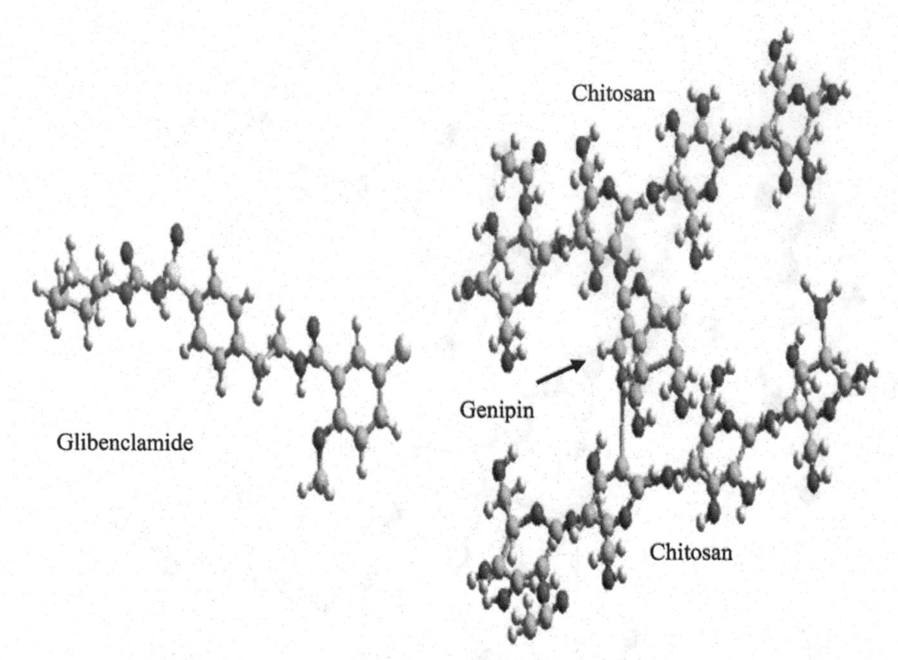

Figure 1.9 Structure of the polymer matrix with glibenclamide.

Table 1.1 QSAR properties and energy (G)

		Values	
Properties	Units	AM1	Experim.
Superficial area	Å^2	796.93	
Volume	Å^3	1,351.38	
Mass	amu	494.01	494.004
Log P		3.50	4.7
E	Kcal/mol	–6,277.66	
ΔHf	Kcal/mol	–156.1536	
A	Kcal/mol	–141,173.2	

1.3.1.2 FTIR

The main assignments of FTIR molecule glibenclamide include the following: In the range of 3494–3468 cm^{-1}, a vibrational mode voltage between the NH amide bonds, which is reported in a rank of 4000, is observed as 2900 cm^{-1} as published by Rubinson K. [17]. However, for the particular case of a secondary amide, it is considered belonging to a signal around 3300 cm^{-1} [18].

Similarly, in the range of 3249–3133 cm^{-1}, signals belonging to the vibrational mode voltage of the CH bond, which is reported in the range of 4000–2900 cm^{-1}, are seen, and the signal belonging to the voltage CH located on the benzene ring is located at 3175 cm^{-1}, reporting in the range of 3150–3050 cm^{-1} [17].

Likewise, the band in the region of 1972 cm^{-1} was assigned to the C=O bond, which is experimentally between 2000 and 1500 cm^{-1}, when it is the vibrational mode voltage, being a lowest frequency signal compared to the C=O of aldehydes and ketones because of its resonance with the nitrogen atom [17], and finally has the signal belonging to S=O sulfoamide a wave number of 1102 cm^{-1}, which is reported in the range of 1200–1140 cm^{-1} in the vibrational mode voltage [19].

1.3.1.3 Electrostatic potential map

The electrostatic potential map calculated for the molecule glibenclamide shows a large electron density, which means a high existence of electrons and prone molecule an electrophilic attack (Figure 1.10) and which can be attributed to the similar electronegativity of carbon, oxygen, and nitrogen, a pair of free electrons exist in the molecule and atomic charges present where most are small, even negative values, and where the positive zone seen in the sulfur atom with a load of 2877 eV. It is recalled that the negative areas are shown in red and positive in blue. It should be mentioned that Spartan if areas with higher electron density due to the range of potential values differ.

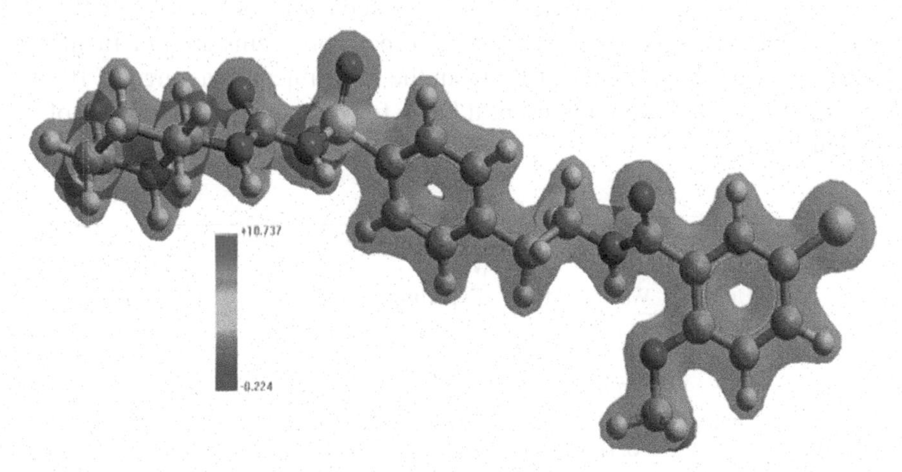

Figure 1.10 Electrostatic potential map of glibenclamide.

1.3.2 Structural Analysis of the Water Molecule and G/A

1.3.2.1 QSAR properties and energy

According to the analyses performed in the molecule of glibenclamide, a minimum water solubility is observed; however, the ratio of the concentration of G/A (glibenclamide/water) is obtained by taking the Log P of 3.5 and Log S of –5.4 mol/L values, which were calculated by the AM1 method and Equation 1.19, respectively, and a G/A concentration of 1.9 mg/L was obtained, confirming the low solubility of glibenclamide in this solvent. It is noteworthy that the software 2.1v ALOGPS calculates a solubility of 2.6 mg/L [16].

Table 1.2 shows the QSAR properties and energy present in the water molecule and the 1:1 ratio. Where in the case of G/A slightly larger than the molecule glibenclamide (G), dimensions and higher binding energy due to the addition of O–H bonds are observed. In addition, a negative free energy and spontaneity indicating this union and a heat of formation higher than glibenclamide pure molecule, generating a higher energy release, occur. This is consistent with that reported by Kurt C. Rolle showing a heat of formation of water molecule –57.79 Kcal/mol [20].

1.3.2.2 FTIR

In the assignments, you can see the three vibrational modes that have the molecule by having a linear and symmetrical structure. The values obtained by the semiempirical AM1 method are presented for comparative purposes and experimental values are reported and calculated by the program HCVIBS consulted in the literature. The data shown in both programs are practically the same; however, these values differ from the experiment in a range of 10–30%. Despite this, they may represent a basis for the general knowledge of the molecule.

Table 1.2 QSAR and energy

		Values			
Property	Units	Water (AM1)	Water (Experim.)	A/G (AM1) (Experim.)	A/G (Experim.)
Superficial area	$Å^2$	114		823.15	
Volume	$Å^3$	117.63		1,413.02	
Mass	amu	18.02	18.015	512.02	512.054
Log P		–0.51		4.40	
E	Kcal/mol	–223.09	–219.02	–6,506.08	
ΔHf	Kcal/mol	– 59.26	–57.79	–216.943	
A	Kcal/mol	–8,038.22		–149,213	

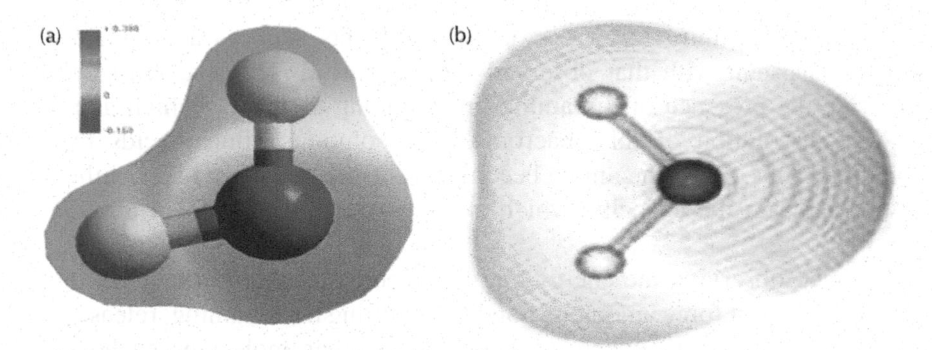

Figure 1.11 Electrostatic potential of water: (a) HyperChem and (b) Spartan software.

1.3.2.3 Electrostatic potential map

In Figure 1.11 the electrostatic potential of the water molecule, where, in paragraph (a) have the electrostatic potential of the water molecule calculated with HyperChem, where shown is noted that the electrophilic area or higher electron density in the area of oxygen atom, this is consistent with the fact that oxygen more electronegative, polarize to him most electronics density while nucleophilic area or the positive part of the molecule in the area of the oxygen atoms. Between these two areas you can see a small green ring indicating a neutral zone within the molecule. Subparagraph (b) the same molecule, taken from the literature and conducted with the SPARTAN program is observed [21].

Purple indicates the electrophilic area and green indicates the nucleophilic area. It also features a small clear ring that divides these two areas and indicates the neutral zone, so it can be seen that the calculation through these two programs is very similar.

1.3.3 Structural Analysis of Chitosan

1.3.3.1 QSAR properties and energy

In this structural analysis, it is considered a deacetylation of 75% reported by several research groups such as Trung Trang S. et al. [22], who obtained excellent results in the properties of the matrix during water absorption property because this directly influences the solubility thereof, in addition to considering existing commercial presentations of chitosan. This molecule is represented by the repetition of four monomers containing amino groups and 3 available an acetyl group. The value found in the Log P of –6.97

indicates a high affinity for water or polar solvents such as acetic acid, which is consistent with that observed by Lárez et al. [23] and reasserts considering the presence of the amino groups along the chain, which generates the dissolution of this macromolecule solutions diluted acids by protonation. Thus, when the amine becomes positively charged, hydrophilic chitosan increases the capacity, which is corroborated by a pKa of amino group of 6.5.

Moreover, observing the energy that has, you can say that it is a stable molecule with great force among its bonds and, during their training, releases large amounts of energy when compared with other molecules such as glibenclamide; also, it has a spontaneous character.

1.3.3.2 FTIR

The principal signals observed in the infrared spectrum are as follows: the presence of the vibrational mode voltage between the oxygen atoms and hydrogen (O–H) in a range of 3493–3451 cm^{-1}. This bond is important because the OH group provides hydrophilic properties to chitosan, directly influencing the water absorption capacity. Likewise, it has a signal of 3451 cm^{-1} belonging to the vibrational mode voltage or symmetric stretching of the amine group, which is the most representative, being this which determines the degree of deacetylation of the biopolymer and therefore the difference between chitin and chitosan. A 2000 cm^{-1} absorption signal belongs to the vibrational mode or tension stretching of the carbonyl group (C=O), which is part of the acetyl group of the molecule (–C=OR), which determines as started the amino group, the deacetylation degree, or acetylation of the molecule as well as a signal in the range of 1550–1520 and 1216 cm^{-1} belongs to vibrational bending mode in the carbon oxygen bond (CO) in the area called the fingerprint.

1.3.3.3 Electrostatic potential map

Figure 1.12 shows the electrostatic potential map where you can see a molecule with varied electronic distribution, being susceptible to electrophilic attack areas where the atoms bonded are COH and HNH (amine), specifically being the oxygen atoms and nitrogen the more electronegative. This is consistent with that reported in the literature, which classifies the amine as a nucleophile [24]. While the carbon backbone shows neutral zone, hydrogens, as expected, show the positive and susceptible of a nucleophilic attack area.

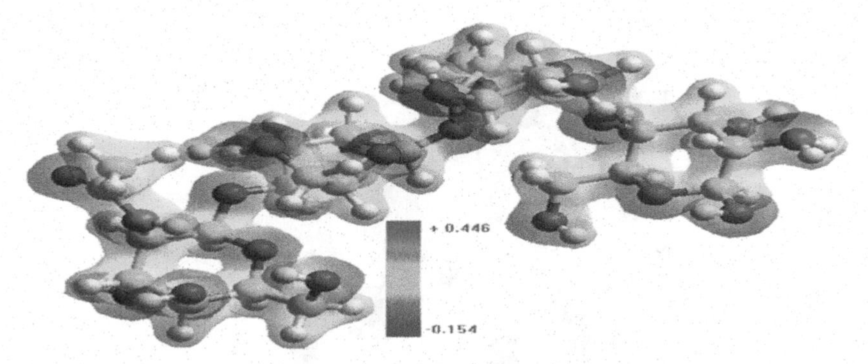

Figure 1.12 Electrostatic potential map of chitosan.

1.3.4 Structural Analysis of Genipin

1.3.4.1 QSAR properties and energy

The values obtained show an exothermic and spontaneous reaction. Full results are shown in Table 1.3. This type of analysis has already been reported by Amani [25].

1.3.4.2 FTIR

Signals in the range of 3502–3465 cm^{-1} belonging to the vibrational mode of OH stretch and signals in the range of 3173–3009 cm^{-1} assigned to the symmetric stretching vibrational mode of binding are observed HCH; signal of 2071 cm^{-1} was attributed to the carbonyl group C=O, as well as signals bending in the area of fingerprint of the CO bonds, CC, and CN mainly.

1.3.4.3 Electrostatic potential map

The electrostatic potential map (Figure 1.13) corresponding to the molecule genipin has a great positive and neutral zone in the hydrogens and carbons,

Table 1.3 Properties of genipin

		Values	
Property	Unit	AM1	Experim. [25]
Superficial area	$Å^2$	416.18	
Volume	$Å^3$	659.81	
Mass	amu	226.23	226.226
Log P		–0.77	
E	Kcal/mol	–3,096.675	
ΔHf	Kcal/mol	–189.66	
A	Kcal/mol	–73,871.34	

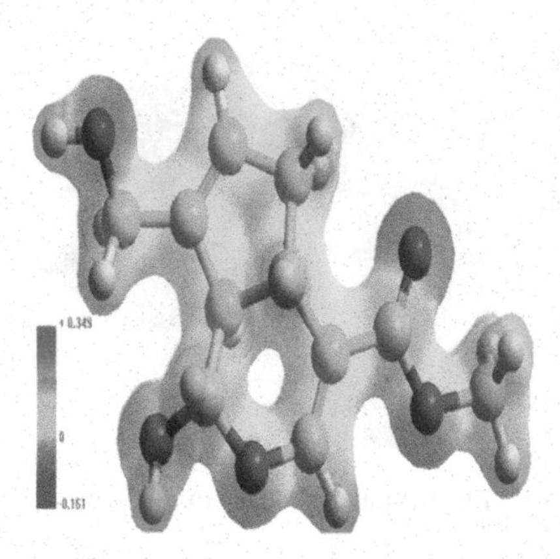

Figure 1.13 Electrostatic potential maps of genipin.

respectively, observing small negative areas in the oxygens, where are the largest electronic density and the increased possibility of an attack electrophilic. It is worth mentioning that the electrostatic cloud shown on the molecule represents the area of influence of the electrons.

1.3.5 Cross-Linking: Chitosan/Genipin (C/Ge)

1.3.5.1 QSAR properties and energy

According to the properties calculated and presented in Table 1.4, it is observed that cross-linking between chitosan and genipin (C/Ge) is carried out showing good affinity with water, according to the Log P value of –14.04 having molecule.

Also, it follows that it is exothermic, spontaneous, and slightly less stable and stronger than those of the individual substances by the heat of formation and the binding energy that occurs in the matrix creating bonds reaction.

Table 1.4 Properties of C/Ge

		Q/GE	
Property	Unit	AM1	Experim. [25]
Superficial area	$Å^2$	1,577.01	
Volume	$Å^3$	3,379.54	
Mass	amu	1585.45	1586.106
Log P		–14.04	
E	Kcal/mol	–20,982.734	
ΔHf	Kcal/mol	–1,742.145	
A	Kcal/mol	–542,690.55	

1.3.5.2 Electrostatic potential map

In Figure 1.14 the electrostatic potential map was calculated for cross-linking Q/Ge where, the electron density is observed homogeneously distributed seen, a slight lack of electrons can be seen in the hydrogen atoms, and observed that the greatest concentration of these are found in amines and oxygen, which have an average of –3.30 eV load, whereas hydrogen has a load 0.14 eV on average.

Partial loads change during the creation of the matrix compared to individual molecules, suffering slight changes in the intensity map areas, although the electron density remains almost constant. Thus, the more reactive atoms that could incorporate new molecules are oxygen system in general. The neutral areas are in the carbon chains. Recalling the areas with a red color are areas susceptible to electrophilic attack. It is noteworthy that the biopolymer chitosan covalently bound to the molecule genipin changes the amine group, which can be seen by the creation of the amide group, which gave new features to the molecule.

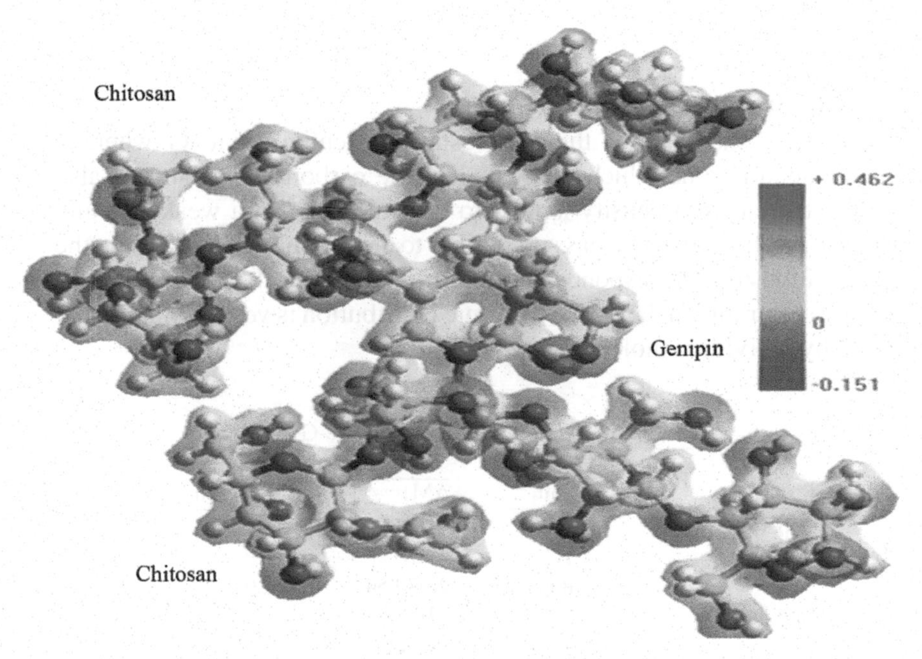

Figure 1.14 Electrostatic potential map of cross-linking: chitosan/genipin.

1.3.6 Adsorption of Glibenclamide in Chitosan/Genipin

1.3.6.1 QSAR properties and energy

In Table 1.5, QSAR properties and different types of energy of the molecule (E, A, ΔHf) are shown, which affords a Log P of −10.87, whereby the strong affinity to water of the Q/Ge-G is corroborated, which influences the free flow of liquid causing swelling of the system. Similarly, the energy values show stability in the molecule and forming source and spontaneous exothermic.

1.3.6.2 FTIR

The main assignments to the signals observed in the vibrational spectrum have vibrational modes of stretching and bending of the binding of OH atoms in a range of 3457–1527 cm^{-1}, the vibrational mode of stretching or tension of carbonyl C=O glibenclamide in the range of 2013–1976 cm^{-1}, and C=C to 1842 cm^{-1} in the genipin. Similarly, the vibrational mode voltage O=S=O of glibenclamide to 1106 cm^{-1} and existing bonds are shown in the fingerprint.

1.3.6.3 Electrostatic potential map

In Figure 1.15, the map of the electrostatic potential Q/Ge–G where a significant neutral zone which is located in the carbon chains shown also shows that the electron density is next to the oxygens and a weaker amino groups (amine), being these areas susceptible to electrophilic attack, while the slightly positive zone corresponds to the hydrogens and with greater intensity in the bond sulfur glibenclamide. Electronic distribution is very similar to that presented by individual molecules.

Table 1.5 QSAR and energy of Q/Ge-G

		Values	
Property	Unit	AM1	Experim.
Superficial area	$Å^2$	2,365.02	
Volume	$Å^3$	4729	
Mass	amu	2,079.54	2,079.49
Log P		−10.87	
E	Kcal/mol	−27,255.86	
ΔHf	Kcal/mol	−1893.760	
A	Kcal/mol	−683,859	

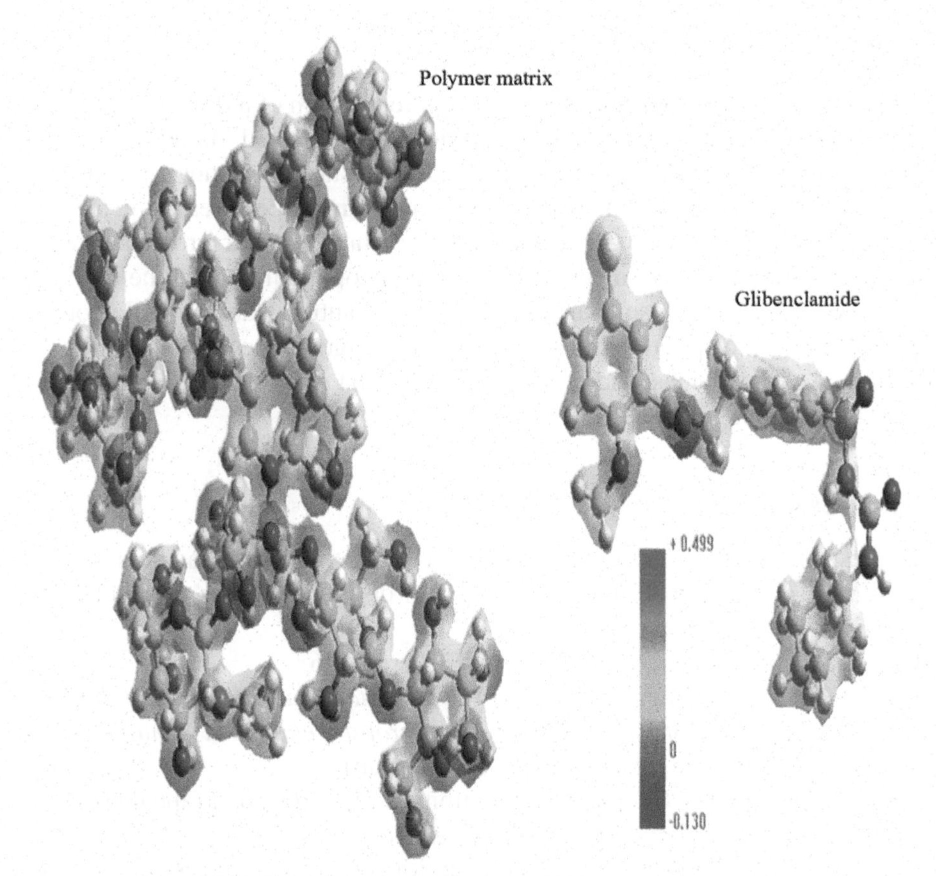

Figure 1.15 Electrostatic potential map of Q/Ge–G.

1.4 Conclusions

Through computational chemistry, it is possible to determine molecular structural properties, having two types of models for this, based on MM and so-called AB–initio which is based on QM; however, due to the computational expense that it accompanies, it is often more practical and functional use semiempirical models that originate in the AB–initio, but include experimental values, avoiding robust calculations and achieving a lower computational cost.

However, it is necessary to mention that the type of model and parameter used must be carefully selected according to the type of system to be analyzed for more accurate and reliable results. Finally, by the way of recommendation, within the semiempirical AM1, PM3, and PM6 models, the most current and certain improvements are enhanced.

Acknowledgments

There are people who contribute greatly to the realization of a dream, a goal. These people are rarely aware of their mission accomplished, do not expect anything in return, do it with great gusto and enthusiasm, and not ask to be recognized. However, gratitude is one of the most human gestures there, and now, this project with such gusto was made crystallized, and irremediably arises the need to give my sincere thanks to the people who paved the way. Thanks to my husband Jose Manuel Hdez., my beautiful baby, and Hector Manuel Hdez. by inspiration. Thanks to my colleagues, the Technological Institute of Aguascalientes, and of course Dra. Norma Aurea Rangel Vázquez for his supervision, guidance, and support.

References

[1] Font M. (2005). Mecánica cuántica. Presentación. Facultad de química orgánica y farmacéutica, sección de modelización molecular. Universidad de Navarra.

[2] https://cuentoscuanticos. Consultado Febrero 2016.

[3] Rangel V. N., and Rodríguez F. (2014). *Computational chemistry applied in the analyses of chitosan/polivinylpyrrolidone/mimosa tenuiflora*, Chapter 1. (México: Science Publishing Group).

[4] King, R. B. (1983). Chemical applications of topology and graph theory. *Studies in Physical and Theoretical Chemistry*, Elsevier.

[5] Sexton, W. A. (2006). Chemical constitution and biological activity. *Am. Pharm. Assoc.* 39, 597–597.

[6] Johnson, A. M., and Maggiora, G. M. (1990). *Concepts and applications of molecular similarity*. (New York: John Willey & Sons).

[7] Castillero E. L. (2013). Espectroscopía Raman de moléculas de simetría tetraédrica, cálculos mecano-cuánticos. Triple enlace, Química.

[8] Zorrilla C. D. (2001). Cálculo teórico de propiedades moleculares mediante bases no estándar. Tesis Doctoral, Universidad de Cádiz, España.

[9] http://www.ehowenespanol.com/ftir-sobre_156331/ consultado mayo 2014.

[10] Manual de Usuario "Hyperchem".

[11] Nicholson, J. W. (2006). The Chemistry of Polymers. Artículo científico digital. University of Greenwich.

[12] Muñoz, R. G. (2012). Obtención de Hidrogeles de Quitosano a Partir del Micelio de Aspergillus Níger Para La Liberación Controlada de Fármacos, Tesis de Maestría en Ciencias-Química, Universidad del Valle, Colombia.

[13] Shirai, M. K. (2012). Bioproceso para obtener Quitina y Quitosano. Artículo científico digital, Universidad Autónoma Metropolitana, 2012.

[14] Allen, T., and Shonnard, R. (2002). *Green engineering. Environmentally conscious design of chemical process*. (USA: Prentice Hall PTR).

[15] Expósito, H. R. (2010). Quitosano, un biopolímero con aplicaciones en sistemas de liberación controlada de fármacos. Tesis Doctoral, Universidad compútense de Madrid, España.

[16] Drug Bank data base, open data drug & drug target database. Version 4.1. Programa/Base de datos de fármacos.

[17] Rubinson, K. A., and Rubinson J. F. (2000). Análisis instrumental. Pearson Educación, U.S.A.

[18] www.depa.fquim.unam.mx/amyd/archivero/Tabla_de_IR_20785. Consultado septiembre 2014.

[19] http://blocs.xtec.cat/1213bat2/files/2013/04/3.-espectroscop%C3%8DA-IR.pdf. Consultado junio 2014.

[20] Kurt, C. R. (2006). *Termodinámica*. (México: Prentice Hall).

[21] https://es.scribd.com/doc/222277746/Electrostatic-A. Consultado agosto 2014.

[22] Trung, T. Si, and Thein-Ha, W. (2006). Functional characteristics of shrimp chitosan and its membranes as affected by the degree of deacetylation. Elsevier, 97, 659–663.

[23] Lárez, V. C. Algunos usos del quitosano en sistemas acuosos. Revista Iberoamericana de Polímeros, 2003, 91.

[24] Bailey, P. S., and Bailey, C. A. (1998). Química orgánica: Conceptos y aplicaciones. (U.S.A: Pearson Educación).

[25] Amani, E., Mayas, A., Nidal, Q., and Asim F. (2011). Chitosan–sodium lauryl sulfate nanoparticles as a carrier system for the in vivo delivery of oral insulin. AAPS *Pharma. SciTech*, 12, 958–964.

2

Application of Quantum Models in Molecular Analysis

Norma-Aurea Rangel-Vázquez[1]
and Nancy-Liliana Delgadillo-Armendariz[2]

[1]PCC, Aguascalientes, Mexico
[2]Chemical Engineering Department, PhD in Engineering Sciences, Technological Institute of Aguascalientes, Mexico

Abstract

Molecular simulations are of great fundamental and applied importance in science and industry and provide fertile ground for the continuing development of theoretical methods. Understanding how atomic-level interactions, structure, and dynamics in molecular materials dictate such macroscopic properties is essential for rational material design. In the recent years, carbohydrate and biodegradable polymers have been extensively used to develop the controlled release formulations of drugs having short plasma life. Controlled delivery systems provide an alternative approach to regulating the bioavailability of therapeutic agents. Biocompatible, biodegradable hydrogels have been designed using natural polymers that are susceptible to enzymatic degradation, or using synthetic polymers that possess hydrolyzable moieties. Of these, hydrogels using the natural polymer, chitosan, have received a great deal of attention due to their well–documented biocompatibility, low toxicity, and degradability by human enzymes. In the present, project has selected the AM1 method in the HyperChem software, mainly for the excellent results with organic molecules as well as the accuracy and speed of the calculations.

Keywords: Metformin, chitosan, genipin, matrix.

2.1 Introduction

2.1.1 Election of the Quantum Model

Among the models based on quantum mechanics, there are mainly Hartree–Fock models, density functional models, LMP2, PM2, and semiempirical model. However, not all are used interchangeably, but if the most accurate in the calculation of the equilibrium results are sought, the geometries of transition state conformations and chemical reactions, it is necessary to choose the most appropriate model for the characteristics of system, then Table 2.1 shows a general guide based on an extensive series of comparisons between quantum models and models of molecular mechanics, provided by the computer suite Spartan, where the letter G means they are good, C is good, but with precautions in the several applications and, P is poor or not recommended.

From Table 2.1, it can be observed that for calculating geometry balance, semiempirical methods are good, in contrast to the thermochemical calculations, covering molecular energies as heats of formation, where the results shed are not as good as calculated by the density functional methods or recipes energy (T1). However, the computational cost is higher, so it must be established very well the project objectives and cost-benefit in time and accuracy, to choose the most appropriate model.

2.1.1.1 Choice of model in basic molecular properties

a) Geometries

In general, all models offer good equilibrium geometries of organic molecules. Semiempirical models rarely show poor geometries and should

Table 2.1 Comparison of quantum models

Property	Molecular Mechanics	Semiempirical	Hartree–Fock	DFT	MP2	T1
Geometry (organic)	C	G	G	G	G	G
Geometry (metals)	–	G	P	G	P	–
Geometry (transition state)	–	C	G	G	G	–
Conformation	G	P	G	G	G	G
Thermochemical (general)	–	P	C	G	G	G
Thermochemical (iso**desmic**)	–	P	G	C	G	G
Computational cost	Low ——————> High					

be used where possible to the start of the calculations, especially considering the little computational expense involved. The Hartree–Fock models do not provide reliable geometries compounds incorporating transition metals, but PM3 semiempirical model or functional density models offer good results. In the case of MP2, it shows good geometries for some transition metals, but poor for others.

b) Conformation Energies

Hartree–Fock and MP2 models with larger basis 6–31G* and density functional models generally provide good results in energy conformation of organic compounds. MP2 and density functional based B3YLP are the most reliable. It must be noted that the difference in the computational expense including small systems can be handy to use the B3YLP models.

c) Reaction Energies

MP2 and functional density models are usually more reliable than the Hartree–Fock models to describe the energies of non-isodesmic reactions, except for reactions in which it is explicit breaking or bond formation, while on the other hand, the Hartree–Fock models with bases 6–31 G* and above also provide acceptable results. Consider that the largest bases 6–31 G* generally provide similar reaction energies to this base. In the case of isodesmic reactions, the Hartree–Fock and MP2 are models that provide better results, while the functional density should only be applied with caution. Finally, the semiempirical methods are not good for the calculation of this property in any reaction, including isodesmic processes.

2.1.1.2 Election model according to their origin

- Molecular mechanics models

These models are used for equilibrium geometries and conformations. They are very practical for the little computational expense models.

- Semiempirical models

They are excellent to determine equilibrium geometries of large molecules, where the Hartree–Fock, MP2, and density functional models can be prohibited. They also can be used without problem to determine geometries of transition states where the computational expense of Hartree–Fock, MP2, and density functional does not make them practical. And perhaps, most important is the geometry optimization in balance and transition states of organometallic metals and inorganic transition, where the Hartree–Fock models are known for throwing poor results, and where the computational expense of PM2 and functional density not allowing to use.

In contrast, it is necessary to mention that the semiempirical models are not good at calculating reaction energies including isodesmic processes and calculations of conformational energy differences.

- Hartree–Fock model

They are used to determine equilibrium structures and transition states systems midsize organic character and major groups of inorganic molecules, where the increase in accuracy of the results that cast semiempirical models is required and where the computational expense of MP2 and functional density makes them impractical. It is also used in the calculation of reaction energies (except those where the breaks and formations are explicit bonds) where semiempirical methods give poor results and where the computational expense of MP2 and density functional prohibits their use. This model is not recommended for calculating the energy of reaction in which there is a break explicitly or bond formation and calculation of energies of activation absolute. Finally, it is used to calculate neither equilibrium structures nor transition states for transition metal inorganic and organometallic molecules.

- MP2 and density functional model

These models are used to describe properties thermochemical reactions which are explicit and formations ruptures chemical bonds, and for calculating absolute activation energies. The local density models do not provide acceptable results, but the other functional density models make the analysis for reactions with breaking or bond formation explicitly. Furthermore, in practice, the MP2 models can only be used for relatively small molecules as functional density models are similar to the Hartree–Fock models in terms of computational cost when the system size is medium.

A density functional model is widely used for the calculations of large molecules where the computational expense makes it impractical MP2, and where a Hartree–Fock and a semiempirical method are not sufficient for the required accuracy. Functional models used the density equally to calculate inorganic or organometallic systems, where a Hartree–Fock is insufficient for the accuracy of the results, as in thermochemical calculations and activation energies absolute, particularly in systems where the breaking or bond formation is explicit.

Finally, it should be noted that the computational expense in the choice of model is very important because, as an example, using structure equilibrium, a Hartree–Fock, may represent twice the time used compared to a semiempirical model (e.g., AM1). Or, if a Hartree–Fock used with a base 3–21 G*, it may represent a computational cost 5 to 10 times greater than a base EDF 3–21 G. A density functional 1 represents only a small difference in computational expense compared to a Hartree–Fock calculation with a similar base for a system of small or medium size, and less costly for a large system, while a calculation B3YLP density functional is about 50% more than a Hartree–Fock taken. This applies for calculating geometry optimization and energy calculations. The MP2 is more time-consuming than any Hartree–Fock or functional density, and its use in practice is severely limited.

2.1.1.3 Choice of a semiempirical model

Within the semiempirical methods, the method has proved best heats of formation, which is the recently developed PM6 (parametric method 6) more. But not only it has the advantage in this calculation, but in many others. FP6 is a reconfiguration of PM3 (parametric method 3), which uses theoretical and experimental data of about 9000 compounds. By way of comparison, it can be mentioned that for the MNDO method (moderate neglect of differential overlap), only 39 compounds, in where for the AM1 (Austin model 1) about 200 and the PM3 more or least 500.

It should be noted that the use of semiempirical method called PM7, however, is still not as popular and only found in newly created computer suites. Thus, in Table 2.2 the average error of some methods compared to PM6 for some analyzed structural properties, ie, that in most cases the PM6 model retains a degree of less than other models error is observed.

Similarly, studies realized with different semiempirical models indicate that the PM6 model provides heat more accurate and better geometries training, best describes the hydrogen bond, and is configured for the following atoms: H, He, Li, Be, B, C, N, O, F, Ne, Na, Mg, Al, Si, P, S, Cl, Ar, K, Ca, Sc, Ti, V, Cr, Mn, Fe, Co, Ni, Cu, Zn, Ga, Ge, As, Se, Br, Kr, Rb, Sr, Y, Zr, Nb, Mo, Tc, Ru, Rh, Pd, Ag, Cd, In, Sn, Sb, Te, I, Xe, Cs, Ba, La, Lu, Hf, Ta, W, Rh, Os, Ir, Pt, Au, Hg, Th, Pb, and Bi; while the PM3 model is parameterized to following atoms: H, Li, Be, B, C, N, O, F, Na, Mg, Al, Si, P, S, Cl, Ca, Zn,

Table 2.2 Comparison of computational methods errors

Properties	PM6	PM5	PM3	AM1	Units
Heat of formation ($\Delta H°f$)	8.01	22.19	18.2	22.86	Kcal/mol
Bond lengths	0.091	0.123	0.104	0.13	Angstroms
Angles	7.86	9.55	8.5	8.77	Grades
Dipoles	0.85	1.12	0.72	0.67	Debye
Ionization potential (I.P.s)	0.5	0.5	0.68	0.63	eV

Ga, Ge, As, Se, Br, Cd, In, Sn, Sb, Te, I, Hg, Tl, Pb and, Bi. Finally, the AM1 model to: H, C, N, O, P, S, F, Cl, Br, and I.

However, like any other semiempirical model and including any method of computational chemistry, PM6 shows obvious errors or inconsistencies in their results under certain conditions, for example, the type of molecule studied, the types of bonds present, or the state of matter, among others.

So, you have to, PM6 not accurately predicted vibrational frequencies, mainly in the area of the OH bonds. The model can predict the entropy S and heat capacity Cp, with good accuracy for simple organic species, but slightly lower for more complicated inorganic and organic species. Similarly, some studies conducted have shown that the method PM6, in the case of molecules in the solid state, may incur various errors; however, it remains in continuous improvement [1].

2.2 Application of Quantum Models in the Structural Analysis of a Polymer Matrix for Drug Release

Quantum models in structural analysis can be applied in many systems. In this chapter, a selection of models and computational shown suites, which are applied in a structural analysis of organic molecules. The system is a polymer based on chitosan matrix, whose function is the release of metformin and glibenclamide drugs, which are used in the treatment of type II diabetes mellitus.

Chitosan is a polymer which has been characterized as a polymer with excellent properties for application in the medical area, is compatible with living organisms, and is completely harmless [2]. It comes mainly from chitin, which in turn is extracted from crustacean shells. Use as releasing hydrogels forming drug is common; however, to obtain a stable hydrogel is necessary to use cross-linkers, which are compounds that form a kind of network that keeps firm and in balance.

Among the most common cross-linkers are aldehydes, although genipin is a natural compound that increasingly more used in laboratories, mainly due to their origin and intersects chitosan properly [3]. These hydrogels are used in transdermal patches for the application of drugs in a controlled manner; however, they have many other applications, such as the use of patches for healing burns.

A polymer matrix is then transformed into a hydrogel when the system is the polymer presented in hydrated form with a degree of specific swelling which determines the amount and rate of drug released as well as substances which make possible the stable compound and transport of the active substance.

On the other hand, according to the World Health Organization (WHO), type II diabetes mellitus is a chronic degenerative disease that is increasing worldwide, which according to projections of this organization will be the seventh leading cause of death in 2030. Type II diabetes mellitus is caused by a problem in the way that it makes or uses insulin in the body. Insulin is needed to move blood sugar (glucose) into cells, where it is stored and later used as an energy source. When you suffer from type II diabetes mellitus, fat, liver, and muscle cells do not respond normally to insulin; this is called insulin resistance.

As a result, blood sugar does not enter the cells to be stored for energy. Thus, the sugar cannot enter cells and accumulate abnormally the high levels of insulin in the blood, which is called hyperglycemia. Currently, there are several categories of drugs used to treat this disease when basic care has failed or does not exist; among them are insulin secretagogues (sulfonylureas, megtlinidas, derived from D-phenylalanine), or alpha glucosidase inhibitors [4], as well as insulin. However, both patients and physicians show a rejection of the use of this as an initial phase in the treatment, unless there is an absolute indication, such as pregnancy, allergy oral hypoglycemic, and surgery [5].

Glibenclamide is a hypoglycemic belonging to the group of sulfonylureas. It stimulates the pancreas cells and increases the release of preformed insulin. Also, it increases the sensitivity of peripheral tissues to the action of insulin and decreases hepatic glycogenolysis (breakdown of glycogen) and gluconeogenesis (glycogen formation from glucose food). Their overall effect is a reduction in blood glucose concentration in diabetic patients whose pancreas is able to synthesize insulin [6].

Whereas metformin (Figure 2.1) or metformin hydrochloride biguanide is particularly prescribed in overweight patients as well as children and people with normal weight. Metformin renal function is as effective in reducing the high levels of glucose in the blood such as sulfonylureas, thiazolidinediones, and insulin. Unlike any other antidiabetic, metformin alone does not produce hypoglycemia [7]. Metformin may be administered in conjunction with glibenclamide; the mechanism of action of glibenclamide reduced the levels of blood sugar, stimulating the pancreas to secrete insulin, and metformin, in turn, helps to regulate the amount of glucose in the blood, reduce glucose absorbed through food, and use the own insulin more efficiently [8].

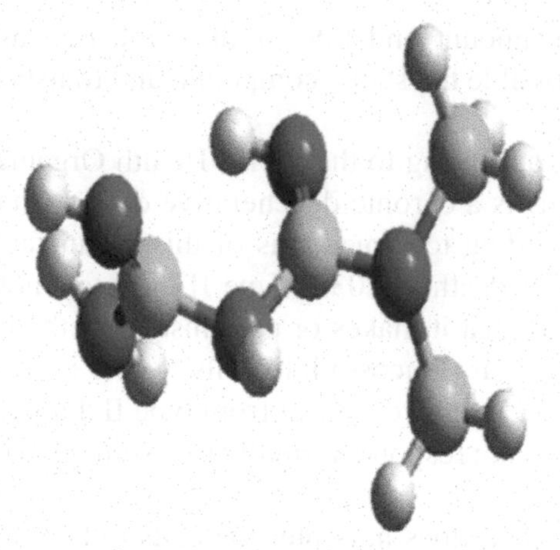

Figure 2.1 Metformin molecule.

2.2.1 Structural Analysis of Metformin

- Physical and chemical properties

Table 2.3 shows the results obtained by HyperChem by means of AM1 model. It should be noted that an energy recipe is an application of various models sequentially on a molecule. The T1 model generally applies in the following recipe or sequence: HF/6–31G* (Hartree–Fock with the basis 6–31G*) Optimization + MP2/6311 + HF/631G * empirical correction.

- Energetic properties

Table 2.4 shows the energy properties of the molecule metformin. Here, you can see an average binding energy of 99.66 Kcal/mol, which means the degree of stability or reactivity which could have the molecule, and that becomes important when compared with other study drugs such as glyburide. This property is important to analyze the reactivity that could have the drug

Table 2.3 Metformin structural properties

Properties	Value
Mass (uma)	129.16
Log P	0.35
Superficial area (Å^2)	265.84
Polarizability	50.31
Volume (Å^3)	138.15
E HOMO (eV)	−8.57
E LUMO (eV)	5.44

Table 2.4 Energy properties of metformin

Properties	Value
Binding energy (E)	99.66
Internal energy (U)	−39, 020.10
Heat of formation (H°f)	61.01
Entropy	0
Gibbs free energy (G°f)	30.92
Helmonts energy	−39, 134.71

passing through the skin, which should not exist. Also, the values of heat of formation (H°f), where it is seen that it is an exothermic reaction (can easily arise without external energy), and the Gibbs free energy of formation (G°f), in where the negative sign indicates that it is an exothermic reaction.

- Electrostatic potential map

Figures 2.2 and 2.3 show the electrostatic potential maps for metformin obtained with Spartan and HyperChem, respectively. These show the areas with higher electron density (red) and consequently where they can attract electrophilic molecules, and areas without electrons (blue), which may be subject to nucleophilic attack. Intermediate color zones contain varying amounts of electrons being located generally neutral zone.

2.2.2 Structural Analysis of Glibenclamide

Currently, they are administered the drugs, metformin and glibenclamide (Figure 2.4). That is why a structural analysis of the molecule of glibenclamide in isolation, shown later in this section, presents a structural analysis of the polymer matrix with both drugs.

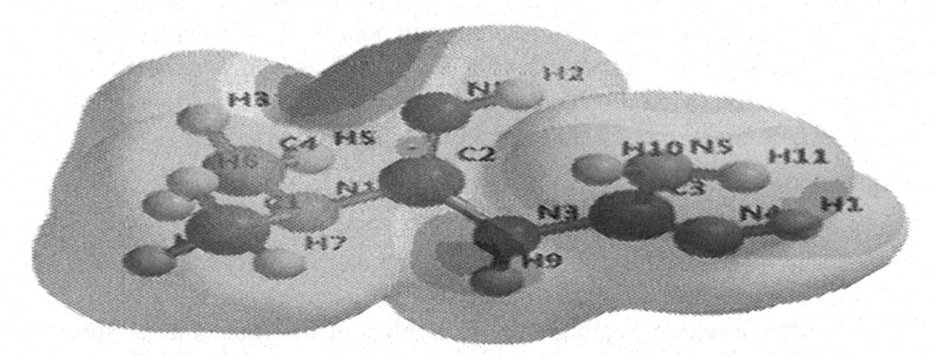

Figure 2.2 Electrostatic potential map of the metformin with Spartan software.

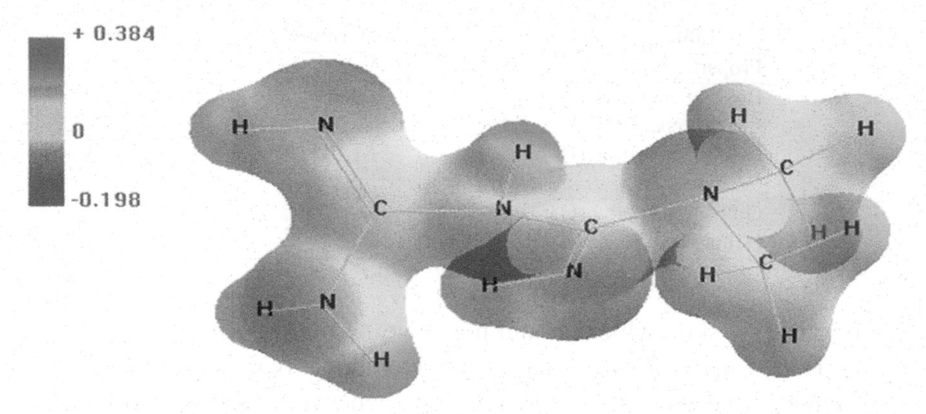

Figure 2.3 Electrostatic potential map of the metformin with HyperChem software.

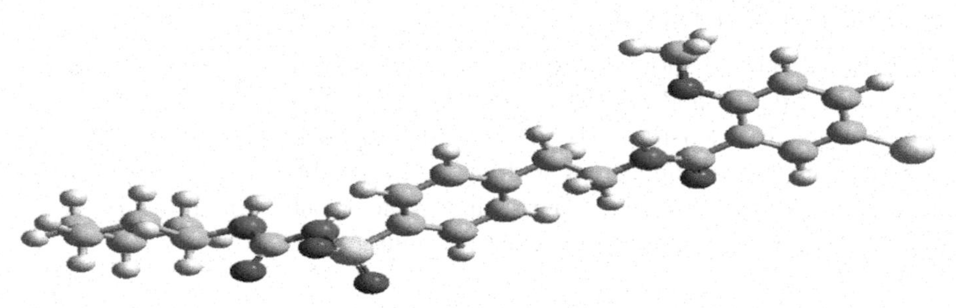

Figure 2.4 Glibenclamide drug molecule.

- Physicochemical properties

Table 2.5 presents the physicochemical properties obtained by HyperChem and AM1 model. It shows a lower percentage of polar area in the case of metformin, a partition coefficient (Log P) positive, which makes a drug of lipophilic nature, indicating a poor dissolution in polar solvents such as water and a large potential in this transfer in lipid media such as skin. Similarly, it is important to mention that a molecule is larger than metformin, with 494 uma calculated evenly by all models and molecular weight reported.

- Energetic properties

Energy values of properties are shown in Table 2.6. It shows average binding energy of 104.62 Kcal/mol, indicating greater stability than average metformin, or what is the same, less reactivity.

This property could indicate greater feasibility of the molecule passing through porous media such as skin, in a more integrated drug metformin. It should be noted that it is intended that the chemicals or drugs that are

Table 2.5 Structural properties of glibenclamide

Properties	Value
Mass (uma)	494.05
Log P	3.5
Superficial area (Å^2)	796.93
Polarizability	76.45
Volume (Å^3)	465.07
E HOMO (eV)	−9.50
E LUMO (eV)	2.31

Table 2.6 Energetic properties of glibenclamide

Properties	Value
Binding energy (E)	104.62
Internal energy (U)	–140,838
Heat of formation (H°f)	–155.15
Entropy	0
Gibbs free energy (G°f)	–144.94
Helmonts energy	–141,173.84

administered transdermally not react in any way in the dermis. On the other hand, according to the sign of the heat of formation and Gibbs free energy of formation, it is a molecule that originates spontaneously and exothermically, so their training is feasible and does not require additional power, indicating compatibility in the elements that form it.

- Electrostatic potential map

Figure 2.5 shows the electrostatic potential map made with the computer suite Spartan. In the electrostatic potential map shown in Figure 2.5, the areas of highest electron density (red zone) and lacking electron areas and, thus having positive charge, attracting nucleophilic molecules (colored areas blue) can be clearly visualized.

2.2.3 Structural Analysis of the Elements of the Polymer Matrix

2.2.3.1 Chitosan

Chitosan (Figure 2.6) is a linear polysaccharide of natural origin, extracted from the shells of crustaceans such as crab or shrimp. Industrially, it is obtained as the product of deacetylation of chitin.

- Physicochemical properties

Table 2.7 presents the structural properties of the chitosan polymer, which is represented by four repeat monomer units and a degree of deacetylation of

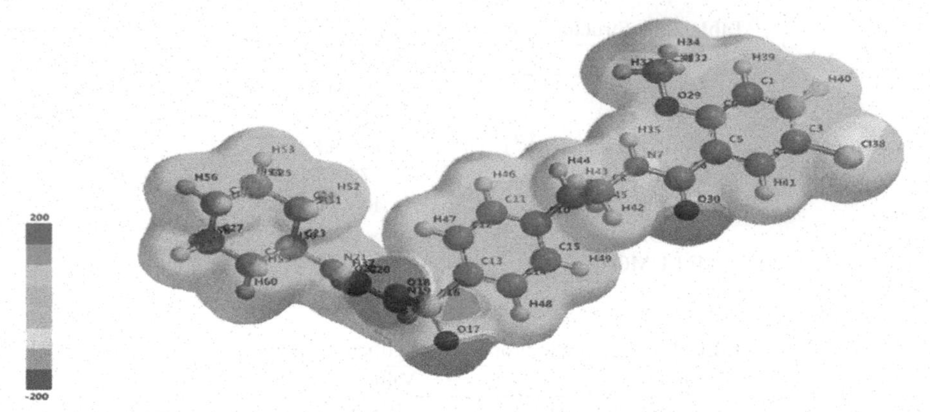

Figure 2.5 Electrostatic potential map of the glibenclamide by Spartan software.

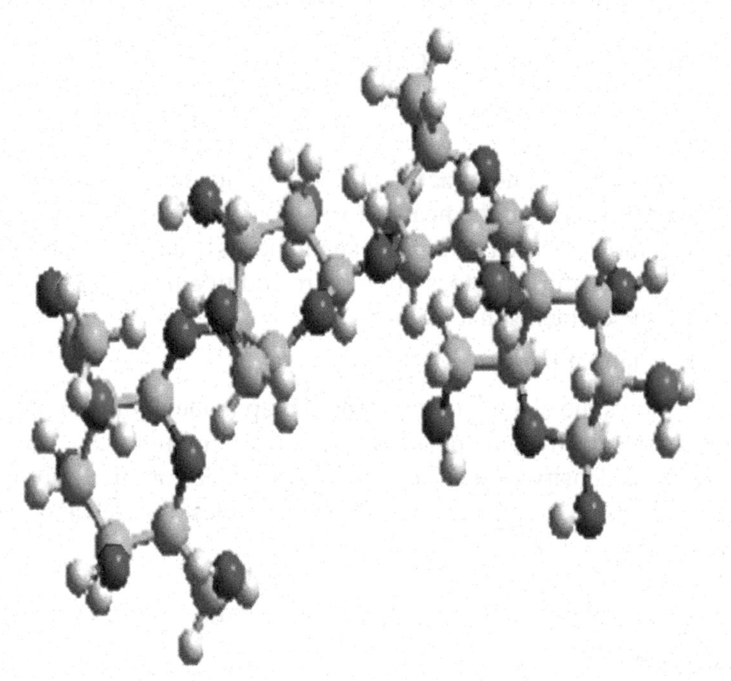

Figure 2.6 Chitosan molecule: 4 repeating units, 75% of deacetylation.

75%, is can be seen the molecular weight with the different computing suites, and total polar surface area which represents 46.96% of the entire molecule. The Log P is negative so that its hydrophilic character is necessary for use as hydrogel.

Table 2.7 Structural properties of chitosan

Properties	Value
Mass (uma)	704.68
Log P	−6.9
Superficial area (Å^2)	873.49
Polarizability	88.09
Volume (Å^3)	626.91
E HOMO (eV)	−8.9
E LUMO (eV)	4.6

Table 2.8 Structural properties of chitosan

Properties	Value
Binding energy (E)	98.48
Internal energy (U)	−243.706
Heat of formation (H°f)	−834
Entropy	0
Gibbs free energy (G°f)	−807.96
Helmonts energy	−243.789

- Energetic properties

Table 2.8 shows the energy properties of chitosan, presenting a negative sign in the heat of formation and Gibbs free energy of formation, which is presented as a molecule with exothermic and spontaneous reaction. In the heat of formation, it can be seen that the most accurate semiempirical method is FP6, which also occurs in the case of the Gibbs free energy of formation when considering the reaction entropy.

- Electrostatic potential map

The electrostatic potential map of chitosan made in Spartan is shown in Figure 2.6. It can be seen that the side of the external hydrogens is the greatest lack of electrons, while the highest density of them is on the side of atoms of internal oxygen. The green and yellow areas are buffer zones with a variable amount of electrons.

2.2.3.2 Genipin

Figure 2.8 shows the structure of genipin/chitosan cross-linking agent of natural origin, extracted from Gardenia jasminoides flower by a hydrolysis process of using the enzyme geniposidio b-glucosidase.

- Physicochemical properties

Table 2.9 indicates the structural properties of the molecule, it can be observed a molecular mass of 226 uma in two suites used are shown, has a

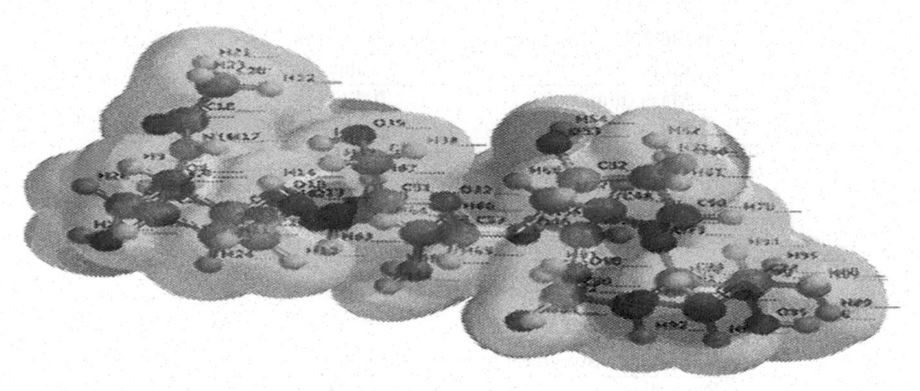

Figure 2.7 Electrostatic potential map of chitosan by Spartan software.

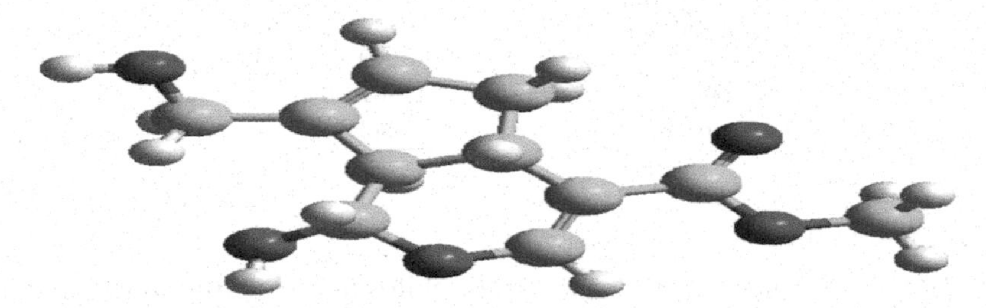

Figure 2.8 Genipin molecule.

log P with negative sign showing its hydrophilicity on both tools, however, the value differs between them showing greater accuracy in the calculated by Spartan. It shows a polar area of 26.9%.

- Energetic properties

According to the values shown in Table 2.10, the heat of formation calculated with the AM1 model used is very similar, being practically the

Table 2.9 Structural properties of genipin

Properties	Value
Mass (uma)	226.23
Log P	−0.25
Superficial area (Å^2)	242.35
Polarizability	57.01
Volume (Å^3)	219.87
E HOMO (eV)	−9.1
E LUMO (eV)	3.4

Table 2.10 Properties energy of genipin

Properties	Value
Binding energy (E)	–106
Internal energy (U)	–73,871.34
Heat of formation (H°*f*)	–189
Entropy	0
Gibbs free energy (G°*f*)	–226.39
Helmonts energy	–73,743.70

same except when energy recipe T1 showing a value –174.46 Kcal/mol is used, and as mentioned before, it is the most reliable method that provides similar data or experimental data. Meanwhile, Gibbs free energy is also very similar among all models, with a value of –211.68 Kcal/mol, using the results of training provided by heat energy T1 recipe. It is noteworthy that these data can be said that is a molecule that originates spontaneously and exothermically form.

- Electrostatic potential map

The electrostatic potential map of the molecule is shown in Figure 2.9. Electron dense areas on the side of the oxygens (red) and main positive side area of hydrogen 15 (blue) can be observed. As shown, Spartan handles as many intermediate zones and divides the zones even more than HyperChem suite.

2.2.3.3 Water

Water (Figure 2.10) as solvent and transport vehicle is a main drug in a polymer matrix or hydrogel, is analyzed also with different models, and suites for the comparison of the values obtained.

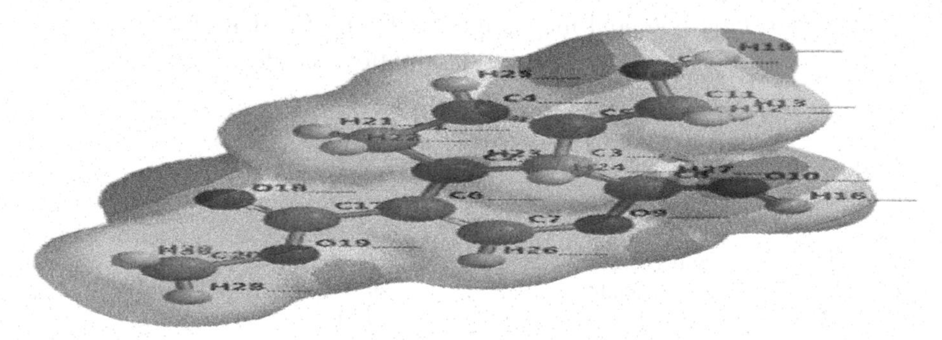

Figure 2.9 Electrostatic potential map of genipin: Suite: Spartan.

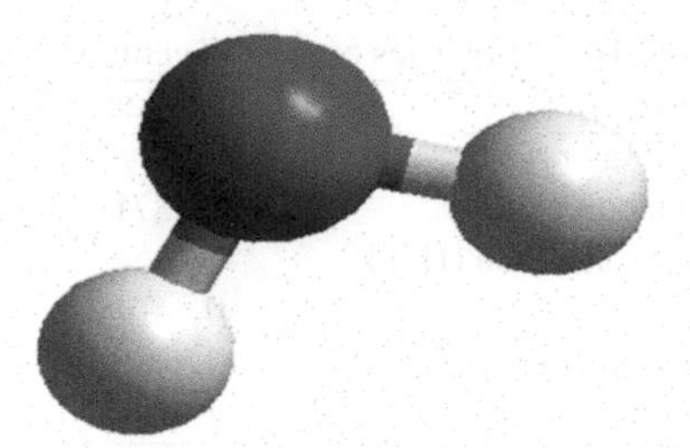

Figure 2.10 Water molecule.

- Physicochemical properties

Table 2.11 shows the structural properties as the mass of 18 uma in all cases, a partition coefficient (Log P) with a negative sign in all models but differ in the numerical value. However, the sign is sufficient to determine the lipophilic nature of the molecule qualitatively. The difference value might have occurred because water is highly polar (polar surface 99.9%), and hydrogen bonds play an important role in their characteristics; thereby, calculating the performing models can differ when considering these qualities.

- Energetic properties

In Table 2.12, the energy properties are observed. Their average energy of binding is –111.51 Kcal/mol, showing one of the most stable analyzed molecules in this study, although it is known that it is also one of the most reactive molecules. The heat of formation and Gibbs free energy of formation have similar values in all models when considering the reaction entropy. And it is in this case when the most accurate values are obtained. With this, you can consider that the creation of the water molecule is exothermic and spontaneous.

- Electrostatic potential map

In the electrostatic potential map (Figure 2.11) by Spartan, they can be observed, in the form of strips, different distributions of electrons having a

Table 2.11 Structural properties of the water molecule

Properties	Value
Mass (uma)	18.02
Log P	–0.382
Superficial area (Å^2)	114.56
Polarizability	39
Volume (Å^3)	19.34
E HOMO (eV)	–12.9
E LUMO (eV)	7.1

Table 2.12 Energetic properties of the water molecule

Properties	Value
Binding energy (E)	–111.50
Internal energy (U)	–8038
Heat of formation (H°*f*)	–59.2
Entropy	0
Gibbs free energy (G°*f*)	–56.23
Helmonts energy	–8.037

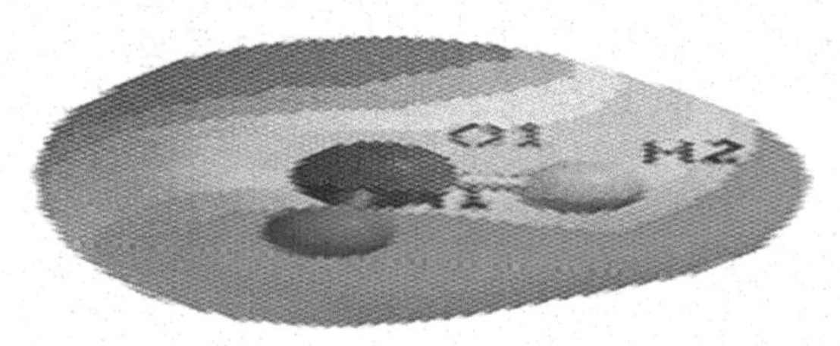

Figure 2.11 Map of the electrostatic potential of water by Spartan software.

high density side of the oxygens (red zone) and ending with a free of them positive zone, in the side of the hydrogens (blue zone). Intermediate or neutral zones are shown with orange, yellow, green, and blue colors according to the amount of electrons existing in them.

2.2.3.4 IR (Infrared)

The infrared spectrum of each of the tested molecules is determined. Table 2.13 allocations and vibration modes of the main components of the polymer matrix individually and comparatively units shown cm^{-1}.

Here, you can see the assignments of the main functional groups and chemical bonds as the HNH bond (amine), of great importance in chitosan since it is the group participating in cross-linking with genipin, and which is located between 3460 and 3999 cm^{-1}.

2.3 System Analysis: Polymer Matrix/Drug

2.3.1 Analysis of Physicochemical and Energy Properties

Table 2.14 presents the physicochemical and energy structural properties of the complete system shown in Figure 2.12, that is, the polymer matrix with glibenclamide and metformin drugs, which would be present without chemical bonding for later release. It is necessary to clarify that the values presented do not consider a system of hydrogel, that is, do not include water molecules. The system is considered as polymer matrix/drug.

Table 2.13 Infrared spectrum of the component molecules of the polymer matrix

Assignments	Vibrational models	Metformin	Glibenclamide	Genipin	Water	Chitosan
O–H	STRETCH	——	3485–3466, 3442	3489–3485, 3465–3440	3583, 3488, 3482, 3463	3640–3250
H–N–H	STRETCH	3493	3462	3465, 3458–3448	3508	——
H–N–H	ESTIRAM.	3460	3460, 3408–3993	3429	3429, 3417, 3408	——
N–H	ESTIRAM.	——	——	3482, 3474, 3447, 3440	3488–3486, 3482–3413	——
C=O	STRETCH	——	2001	2001	——	2000–1500
C–N	FLEXION	——	——	1853	——	1500–600
C=C	STRETCH	——	1850	——	——	2000–1500
C–N–C	STRETCH	17,361,490	17,361,490	——	——	1500–600
O–H	FLEXION	——	1554–1507	1584–1525	1585–1565, 1559–1508	1700–1600
H–C–H	FLEXION	1476–1419	——	——	1489, 1452	1500–600
H–C–H	FLEXION	1383	——	——	1445	1500–600
N–H	FLEXION	1354, 865–838, 715–589	——	1645–1529	1505, 1462	1500–600
H–C–H	FLEXION	1165–1127, 1102–1038	——	1375, 1290	1010	1500–600

Table 2.14 Physicochemical and energetic properties of the polymer matrix/drugs

Properties	Value
Mass (uma)	2,208.71
Log P	–14
Superficial area (Å^2)	2,014
Polarizability	39.5
Volume (Å^3)	2,006.15
E HOMO (eV)	–8.6
E LUMO (eV)	–0.9

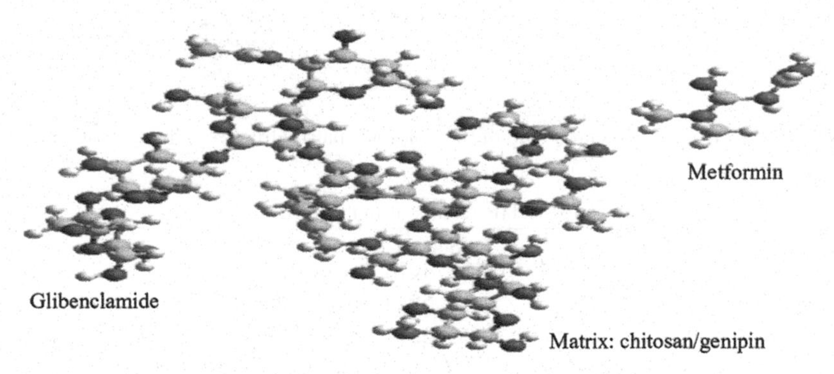

Figure 2.12 Polymer matrix/drugs.

2.3.2 Electrostatic Potential Map

In Figure 2.13, the electrostatic potential map of polymer matrix/drugs is shown. In it can be observed as areas of higher electron density those located in the oxygens of the chitosan chain and positive deficient areas electrons external hydrogens of the polymer matrix, while glibenclamide neutral zones represented color observed green and yellow and a strong negative oxygen zone is adjacent to the sulfur in the molecule. Metformin turn strongly positive sample areas external hydrogens of the molecule. It must be remembered that the intermediate strips are areas where increasing amounts of electrons are represented from deep blue through the clear blue, green, and yellow colors, to bright red representing the highest electron density is located.

2.3.3 IR (Infrared)

Table 2.15 shows the main assignments for this system (polymer matrix/drugs), where you can see little displacement of the bands shown. The model used is PM6, and experimental reference to visualize the proximity to these by computational chemistry data shown at a time.

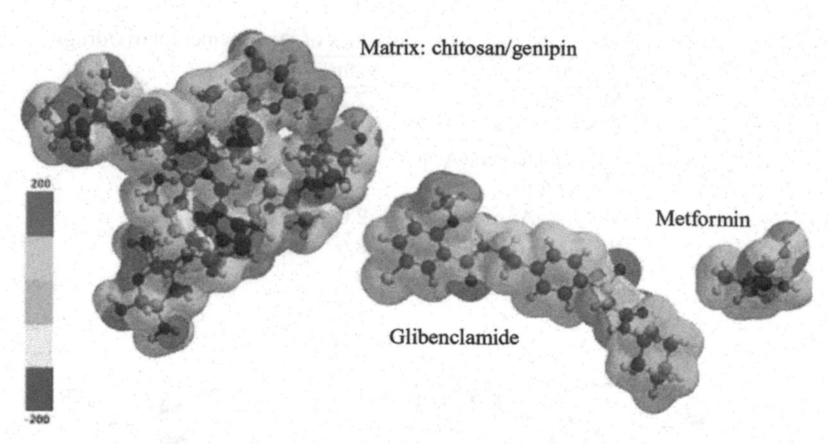

Figure 2.13 Electrostatic potential map of the polymer matrix/drugs, by Spartan software.

Table 2.15 FTIR results of matrix/drugs

Bonds	Frequency (cm^{-1})
N–H stretching	2826, 2810, 2791
C–H stretching	2770–2760, 2717–2712, 2566
O–H stretching	2555–2237
C=O stretching	1789
C–C stretching	17757, 1709, 1644
N–H flexion	1585–1402
C–H flexion	1518, 717, 691, 521
C–C flexion	631–601, 574–501, 382
C–N flexion	514

2.4 Conclusions

No doubt, computational chemistry proves to be a powerful tool for determining molecular properties by the analysis of their chemical structure. And as has been stated, there are various models and computer suites that can be used; however, the choice of these must be based on the type of molecular system to be analyzed and the properties sought to obtain. Thus, it can be concluded that in a general manner, the use of quantum models of semiempirical bases is recommended for optimizations geometries, and the Hartree–Fock or functional density for molecular energies; however, due to the difference of the computational expense, they have different models, using a semiempirical model for this purpose; it is advisable to use PM6 in most cases. Finally, it should be recognized that enegéticas recipes are the most accurate and can be used in small systems but in large molecules or unions of these are not suitable.

Acknowledgments

There are people who contribute greatly to the realization of a dream, a goal. These people are rarely aware of their mission accomplished, do not expect anything in return, do it with great gusto and enthusiasm, and not ask to be recognized. However, gratitude is one of the most human gestures there, and now this project with such gusto was made crystallized, and irremediably arises the need to give my sincere thanks to the people who paved the way. Thanks to my husband Jose Manuel Hdez., my beautiful baby, and Hector Manuel Hdez. by inspiration. Thanks to my colleagues, the Technological Institute of Aguascalientes, and of course Dra. Norma Aurea Rangel Vázquez for his supervision, guidance, and support.

References

[1] http://openmopac.net/Features_of_PM6.html
[2] Font M. Mecánica cuántica. Presentación. Facultad de química orgánica y farmacéutica, sección de modelización molecular. Universidad de Navarra, 2005.
[3] https://cuentoscuanticos Consultado Febrero 2016
[4] Rangel, V. N., & Rodríguez, F. (2014). *Computational chemistry applied in the analyses of chitosan/polivinylpyrrolidone/mimosa tenuiflora*. (México: Science Publishing Group).
[5] King, R. B. (1983). Chemical applications of topology and graph theory. Studies in Physical and Theoretical Chemistry.
[6] Sexton, W. A. (2006). Chemical constitution and biological activity. *Am. Pharma. Assoc.* 39, 597.
[7] Johnson, A. M., and Maggiora, G. M. (1990). Concepts and applications of molecular similarity. (New York: John Willey & Sons).
[8] Castillero E. L. (2013). Espectroscopía Raman de moléculas de simetría tetraédrica, cálculos mecano-cuánticos. Triple enlace, Química.

3

Molecular Analysis of Insulin Through Controlled Adsorption in Hydrogels Based on Chitosan

Norma-Aurea Rangel-Vázquez[1] and Ana-Karen Frías-González[2]

[1]PCC, Aguascalientes, Mexico
[2]Chemical Engineering Department, Master in Science in Chemical Eng. Technological Institute of Aguascalientes, Mexico

Abstract

Computational chemistry comprises areas of chemistry, biology, and physics, coupled with computing, which allows investigation of atoms, molecules, and macromolecules using a computer system. So, the computational chemistry facilitates experimental work and achieves important and difficult or even impossible to obtain by other means data is obtained. For this, there is a wide range of mathematical methods which apply the laws of classical physics. Thus, there are software packages that can solve different types of models to generate data. One of the most efficient and comprehensive programs "HyperChem" is a sophisticated molecular simulation program because of its quality, flexibility, and ease of use, uniting 3D visualization and animation with chemical, molecular quantum mechanical calculations and molecular dynamics.

In this research project, structural analysis of chitosan molecules, genipin, and insulin is performed by the software HyperChem, to obtain physical and chemical properties of the molecules. Similarly the properties for the cross-linking of chitosan with genipin and when added insulin, taking as a basis for the calculation model molecular mechanics, because insulin has in its amino acid structure which are applicable to molecular mechanics were determined, whereby the method used was the AMBER.

The properties obtained were determined the absorption of insulin in chitosan cross-linked with genipin, which can be used as part of a transdermal patch for drug absorption, which is primarily administered by subcutaneous injection. Most patients require insulin, suffer from some side to the post-injection effects such as lipodystrophy, bleeding, or bruising. For this reason and given the need to cover a claim presented by the society in general, this project aims to analyze and predict molecular behavior of insulin adsorption using hydrogels of chitosan.

Keywords: Insulin, adsorption process, matrix, AMBER.

3.1 Introduction

3.1.1 Polymers

Polymers are macromolecules formed by repeated one or more binding molecules called monomers linked together by covalent bonds [1]. Depending on their origin, the polymers may be natural or synthetic. Natural biopolymers or polymers are produced by living organisms, and these are result from only raw materials found in nature. It has recently paid particular attention to natural polymers, because they are biocompatible and biodegradable, so it can be hydrolyzed, and it is removable and non-toxic products [2].

3.1.1.1 Chitosan

Chitosan (Figure 3.1) is a natural polymer derived from chitin, one of the most abundant biopolymers in nature. Chitin is part of the support structure of many living organisms, such as arthropods (crustaceans and insects), mollusks, and fungi. It is also an important byproduct of various industries such as fishing and the beer. Chitin and chitosan are biopolymers which in recent years found in many applications, especially in the food industry and the biotechnology [3].

Obtaining chitosan from chitin by deacetylation is performed the same, leaving the amino group on carbon 2; this process never reaches 100%. That is why the chitosan is a copolymer of 2-acetamido-2-deoxy-β-*D*-glucose and 2-amino-2-deoxy-β-*D*-glucose.

However, when the degree of deacetylation reaches 100%, the polymer is named as chitan, although it is difficult to get this degree. The source and method of obtaining the composition of the chains of chitosan and its size are determined. For this reason, the deacetylation degree and molecular weight

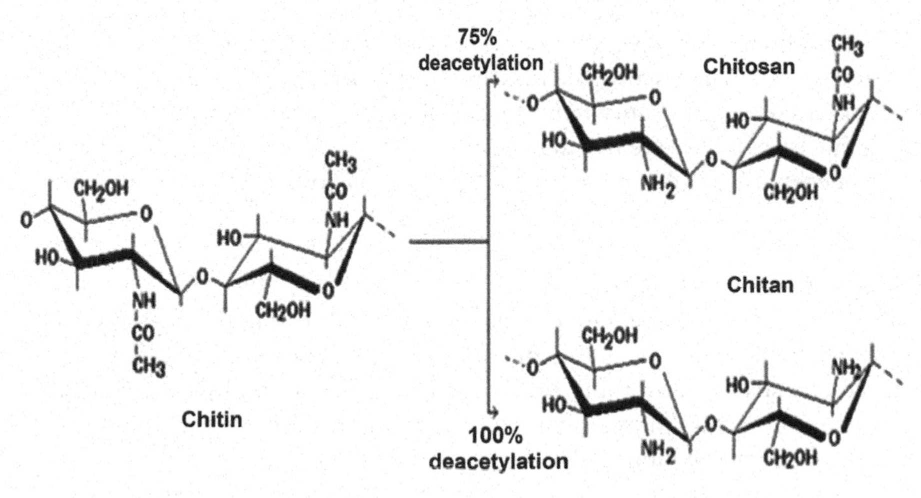

Figure 3.1 Chemical structures of chitin, chitosan and chitan.

average are two parameters of knowledge required for the characterization of this polymer [4]. The main physicochemical properties of chitosan that determine their functional properties are the degree of deacetylation and the average molecular weight, although the crystallinity, water content, ash, and proteins are also physicochemical properties to consider for the application of a specific chitosan [5].

3.1.2 Hydrogels

Hydrogels are cross-linked polymeric materials in three dimensions obtainable by natural or synthetic polymers and monomers. For a material to be classified as hydrogel, it should have the characteristics such as swelling equilibrium and phase transition, and in some cases, it should have sensitivity to medium encompassing.

Thus, with the chitosan, smart hydrogels having the ability to perform a selective swelling against pH due to the amino groups of the polymer chain can be obtained [6]. The hydrogels for their high water content and their ability to retain molecules low molecular weight can be used as controlled release systems, wherein the release occurs by simple diffusion through the polymer matrix swollen and to an external environment.

Cross-linked chitosan hydrogels are classified as ionic and covalent hydrogels. The latter are divided into three groups: chitosan cross-linked with itself, hybrid interpenetrating polymer networks, and polymer networks.

Cross-linked hydrogels have a more compact structure and swell much less compared with the same hydrogel cross-linking [7]. In Figure 3.2, the hydrogel is shown based on chitosan where cross-linking is carried out by scanning electron microscopy, where the main feature observed is the porosity, which can vary depending on the amount, and these dimensions in the synthesis are appreciated.

3.1.2.1 Cross-linking agents

Absorbing systems based on biodegradable polymers need to be cross-linked to modulate their properties and maintain the stability of the matrix and thus fulfill the objective of absorbing drug over the desired time. In the covalent hydrogels, more cross-linking agents used are the genipin and dialdehydes such as glyoxal and glutaraldehyde.

Dialdehydes allow the reaction to occur directly in an aqueous medium without auxiliary molecules (reducing) which can reduce their biocompatibility; however, a disadvantage of using these compounds is their toxicity. The use of genipin has increased by good null toxicity, and cross-linking was observed. The degree of cross-linking is the main parameter in mechanical strength properties, swelling, and diffusion, in addition to providing a porous structure [8].

3.1.2.2 Genipin

The genipin is a naturally occurring compound that is obtained from genipósido from the fruit of *Gardenia jasminoides* and *American Genipa*. These fruits have anti-inflammatory, diuretic, choleretic, and hemostatic properties [9]. Figure 3.3 shows the chemical structure of genipin is presented,

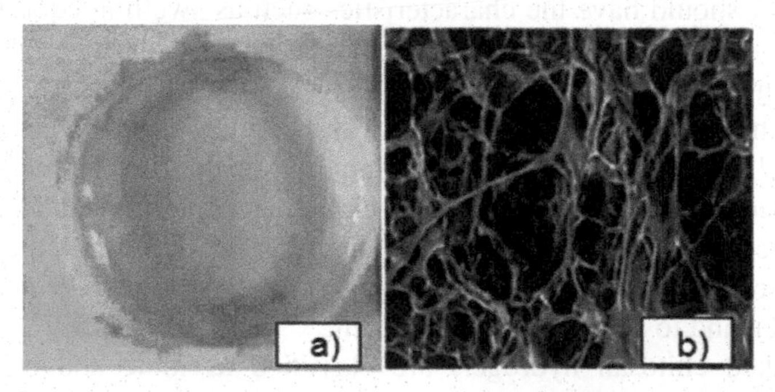

Figure 3.2 Hydrogel about chitosan: (a) swelling the eye and (b) image with scanning electron microscopy (SEM).

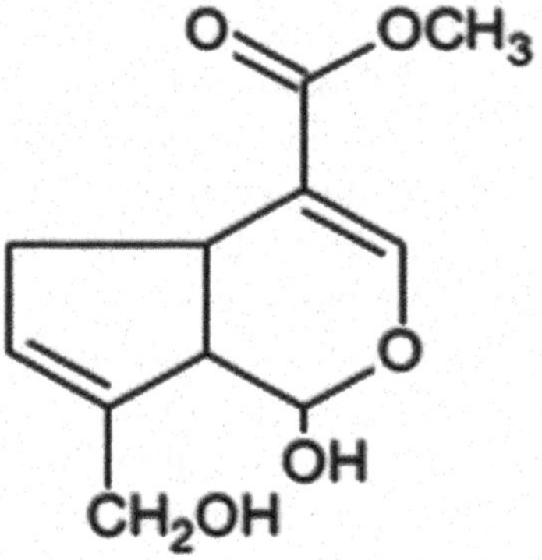

Figure 3.3 Chemical structure of genipin.

which can be seen composition which consists of a methyl ester group attached to a diene formed by two heterocycles and $C_{11}H_{14}O_5$ the chemical formula [10].

During the cross-linking reaction between genipin and chitosan, two separate reactions occurred. The first reaction is a nucleophilic attack by the amino groups of chitosan on carbon 3 of genipin, resulting in the formation of a heterocyclic compound bound to genipin glucosamine residue in chitosan. The second reaction, slower, is a replacement nucloephilic of the ester group of genipin, releasing methanol and forming an amide bond with chitosan [11].

3.1.3 Adsorption of Drugs

Adsorption is the passage of molecules of a drug to the blood. All drugs, except those administered topically in order to produce local effects, must enter into circulation in order to reach the site of action and exert its therapeutic effect [12]. Their number factors influence the adsorption of a drug such as the route of administration, the site of adsorption, solubility, adsorption surface concentration, blood flow, and possible changes in the absorption area. For example, the greater the adsorption will be, the greater the adsorption surface and blood flow, since the area involved in the exchange process is greater and a great vascularization in the area allows quick passage of the drug into the circulation, favoring the transfer process [13].

3.1.3.1 Dermal adsorption

Dermal adsorption is defined as the transport of a compound from the outer surface of the skin on the skin and body, and for chemicals that are absorbed, the vast majority is by passive diffusion following Fick's law. Therefore, the adsorption of the skin means that the compound becomes available systematically. Adsorption into the skin should not be confused with permeation skin, which is the diffusion of a compound in a certain layer of the skin, not with skin penetration is the spread in the deeper layers, as shown in Figure 3.4.

Permeation is the diffusion of a piercing in a certain layer of skin. The subsequent diffusion through this layer represents penetration. The penetration through skin layers, either the site of action or absorption, represents systemic circulation. Therefore, in cutaneous administration, the drug can be absorbed through the skin, and although this administration has as its purpose a local effect, which cannot be guaranteed that no systematic effects occur [14].

3.1.4 Diabetes

According to the World Health Organization (WHO), worldwide, there are more than 347 million people with diabetes, of which it is estimated that in 2012 killed 1.5 million people as a result of excess blood sugar in fasts. Diabetes is a chronic disease that occurs when the pancreas does not produce enough insulin or when the body does not effectively use the insulin it produces [15].

Diabetes is among the leading causes of death in Mexico; in 2012, 6.4 million people reported having been diagnosed with diabetes [16]. One of the major risk factors for developing diabetes is overweight and obesity. In 2012, the leading cause of death of the Mexican population corresponds to

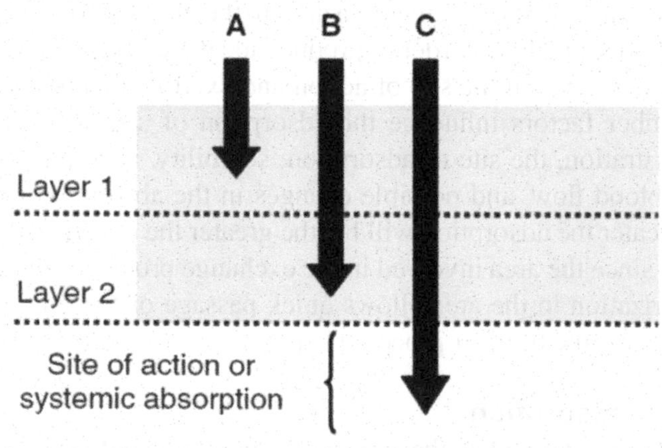

Figure 3.4 Representation of (a) skin permeation, (b) skin penetration, and (c) the skin absorption.

heart disease, which accounts for a fifth of all deaths; followed in descending order, diabetes mellitus, equivalent to 16.6% of total deaths of the victims and 12.2% of them (Figure 3.5), where the states with the highest prevalence are Federal District, Nuevo Leon, Veracruz, Tamaulipas, Durango, and San Luis Potosi.

3.1.4.1 Insulin

Insulin is a peptide hormone of 5.8 KDa and is secreted by the β cells in the pancreatic islets of Langerhans in response to high nutrient levels in the blood [17]. In Figure 3.6, which insulin is a polypeptide composed of two amino acid chains, which in total have 51 amino acids, the A chain is 21 amino acids and the B chain, 30 amino acids shown.

Figure 3.5 Leading causes of death in Mexico.

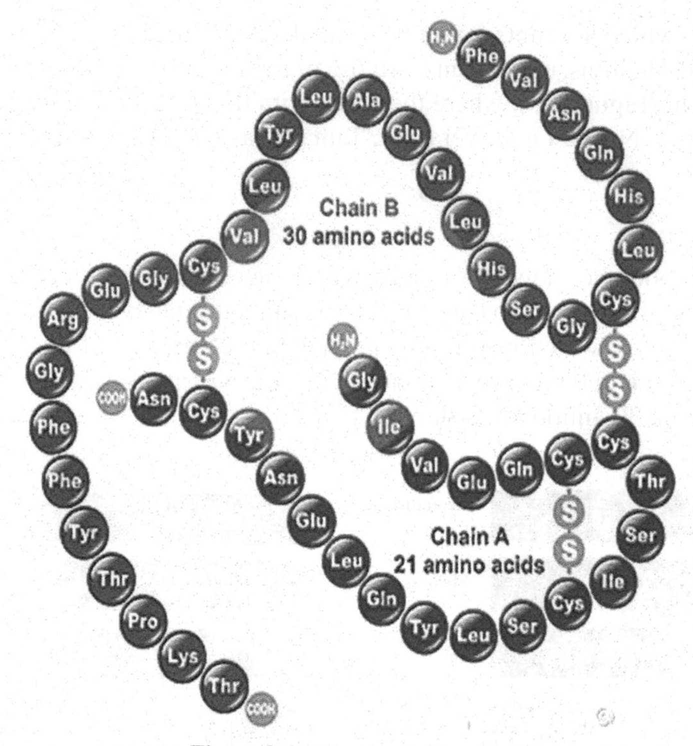

Figure 3.6 Structure of insulin.

Both chains are linked by two disulfide bridges in the amino acid A–7/B–7 and A–6/B–19. Furthermore, the chain also has a disulfide bridge between internal amino acids A–6 A–11 [18].

3.1.5 Computational Chemistry

Computational chemistry is a study of chemical problems at the microscopic level (atomic–molecular) using the equations provided by quantum mechanics (to characterize the electronic structure of atoms and molecules) and statistical mechanics (for macroscopic properties from microscopic constituents). Theoretical chemistry analyzes the mathematical description of chemistry. Computational chemistry refers to its computational implementation for solving corresponding equations. The complexity of solving equation proposals requires addressing the study by introducing approaches and the use of computers [19]. Computational chemistry covers a wide range of mathematical methods which can be divided into two broad categories:

- Molecular mechanics: Apply the laws of classical physics to molecular core without explicitly considering the electrons.
- Quantum mechanics: Based on the Schrödinger equation to describe a molecule with a direct treatment of the electronic structure and is subdivided into two classes according to the treatment performed, semiempirical methods, and *Ab-initio* methods [20].

3.2 Methodology

3.2.1 Determination of Structures Individually

The identification of individual molecules was performed using the software package HyperChem Pro 8.0.6, using the AMBER method. To calculate the properties of the individual components of the molecules, the first thing done was the outline of each by selecting the draw command located in the toolbar.

3.2.2 Calculation of Energy

The energy for each of the molecules was calculated according to the following procedure: the geometry for each of the molecules was optimized by the compute tool, by selecting the geometry optimization. Subsequently, the method of Polak–Ribiere, RMS gradient of 0.05 Kcal/(Å mol), and the option is selected under vacuum.

3.2.3 Obtaining the Partition Coefficient (Log P)

The partition coefficient is selected by choosing the compute/QSAR properties option and the Log P (partition coefficient) option from the menu bar.

3.2.4 Obtaining the Electrostatic Potential Map

After obtaining energy optimization using the AMBER method, you can draw diagrams in three dimensions (3D) Mapped Isosurface. The HyperChem software shows the electrostatic potential a contour plot. This was done by selecting the option compute/plot molecular graphs from the menu bar.

3.2.5 Analysis of the Infrared Spectrum (FTIR)

To determine the wavelengths for the infrared spectrum of molecules, the following route was followed: the compute/vibration rotation analysis option was selected and, subsequently, the vibrational spectrum option. Subsequently, spectra were analyzed in various frequency vibrations by selecting a particular option. When the signal spectrum is selected, the corresponding bond type shows its vibrational mode.

3.3 Results

3.3.1 Structural Analysis of Chitosan

3.3.1.1 Energy optimization and partition coefficient (Log P)

In the structural analysis of the molecule of chitosan, the degree of deacetylation was 75% by Trung et al. who obtained excellent results in the matrix in which higher water absorption was obtained with the percentage considered for chitosan [21]. Table 3.1 shows the values corresponding to the value of the partition coefficient (Log P) and Gibbs energy, respectively. The Log P value of –6.97, indicating a high affinity for water or polar solvents, is consistent with the results observed by Velasquez (2003), which considered that a material with higher charge density generates greater solubility [22]. Moreover, observing the Gibbs free energy which has the chitosan molecule of 85.15 kcal/mol indicates that the molecule tends to react with related compounds not spontaneously.

3.3.1.2 Electrostatic potential map

In Figure 3.7, the potential map with a wide electronic distribution is shown, being susceptible to nucleophilic attack areas (red) of the bonds: COH, HNH, and C=O, specifying the oxygen atoms and nitrogen which are more electronegative. It can be observed that electrophilic attack susceptible areas (green–blue) were appreciated in the hydrogens atoms, and the carbon backbone represents the neutral zone.

3.3.1.3 FTIR

The main assignments observed in the infrared spectrum are shown in Table 3.2. It is posible to appreciate the presence of stretching vibrations between the OH bond between 3766 and 3756 cm^{-1}; ///these bonds are important because OH groups provide hydrophilic properties, influencing

Table 3.1 Properties of chitosan

Properties	AMBER
Gibbs free energy (Kcal/mol)	85.15
Log P	–6.97

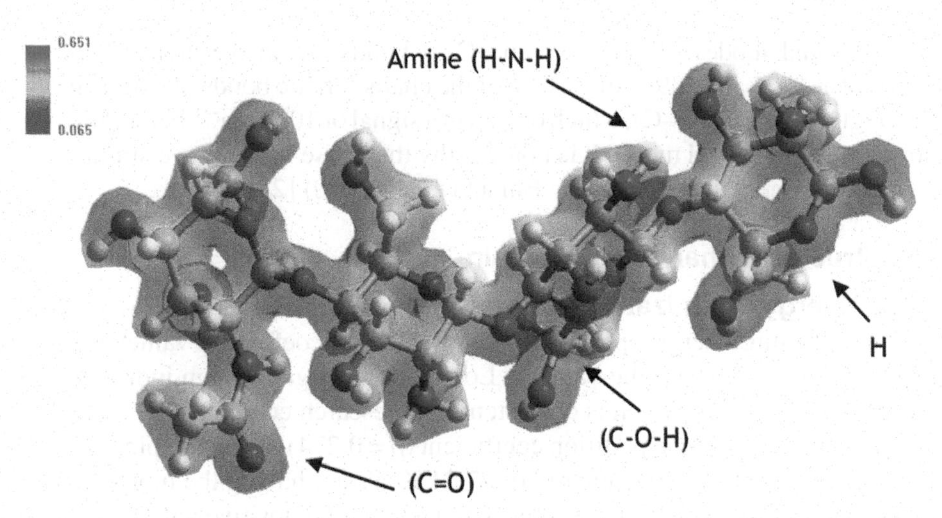

Figure 3.7 Electrostatic potential map of chitosan (75% deacetylation): ● carbon, ● oxygen, ● nitrogen, and ○ hydrogen, respectively.

Table 3.2 FTIR assignments of chitosan (75% deacetylation)

Assignments		AMBER Frequency (cm^{-1})	Experimental Frequency (cm^{-1}) [23]
O–H	Stretching	3766–3756	3414
H–N–H	Symmetric stretching	3414	3215
N–H	Stretching	3282	3500–3300
H–C–H	Asymmetric stretching	3080	2960–2850
C–H	Stretching	2847	2926
C=O	Stretching	1872	1664
H–N–H	Flexion	1673	1574
C–N	Flexion	1475	1313
C–O	Flexion	1396	1074
C–O–C	Flexion	1064	894

water uptake capacity. Similarly, it has a signal of 3414 cm^{-1} belonging to the symmetric stretching vibrational mode of the amine group (H–N–H), which represents the degree of deacetylation of chitosan. Signal at 3080 cm^{-1} corresponding to the asymmetrical stretching vibration between the H–C–H is presented. A 2847 cm^{-1} stretching vibration between the C–H was observed.

Likewise, the signal of 1872 cm^{-1} was attributed to stretching vibrational mode of the carbonyl group (C=O) which is the acetyl group in the molecule.

The vibrational modes of the C–N and C–O bonds are in the area called fingerprint at 1475 and 1396 cm^{-1}. One of the characteristic bonds of chitosan is the O–glycosidic (C–O–C), which occurs in a signal of 1064 cm^{-1} belonging to stretching vibrational mode. It is noteworthy that these values were similar to those experimentally reported by Balanta et al. (2010) [23].

3.3.2 Structural Analysis of Genipin

3.3.2.1 Energy optimization and partition coefficient (Log P)

The corresponding energy optimization and partition coefficient values are shown in Table 3.3, where the value of Log P with a negative sign indicates a hydrophilic character, which is consistent with research conducted by Suk–Yoo et al. who obtained the partition coefficient ($p = 0.274$) of the genipin [24] with respect to the Gibbs free energy of 27.21 Kcal/mol having the molecule follows that it is a non-spontaneous formation due to the positive value of its energy.

3.3.2.2 Electrostatic potential map

The electrostatic potential map of the molecule genipin is shown in Figure 3.8, neutral zones at carbons and small negative areas in the oxygens can be seen where the highest electron density and most are also observed possibility of nucleophilic attack. Likewise, they are positive hydrogen areas which are susceptible to electrophilic attack.

3.3.2.3 FTIR

In Table 3.4, the main infrared spectrum allocations of genipin are observed wherein the presence of stretching vibrations of OH bond is in a signal of 3769 cm^{-1}. Similarly, it has a signal of 3413 cm^{-1} belonging to the vibrational mode of the CH stretching. At a signal from 3064 cm^{-1}, asymmetrical stretching vibration of CH bond is presented. Also, it has a signal of 1914 cm^{-1} belonging to the vibrational mode of the carbonyl group (C=O).

The vibrational mode stretching of C=C was presented at the signal of 1890 cm^{-1}, and at a signal of 1863 cm^{-1}, stretching of genipin ring occurs. Bending signals attributed to CC and CO bonds were seen in the area called fingerprint [25].

Table 3.3 Properties of genipin

Properties	AMBER
Gibbs free energy (Kcal/mol)	27.21
Log P	–0.77

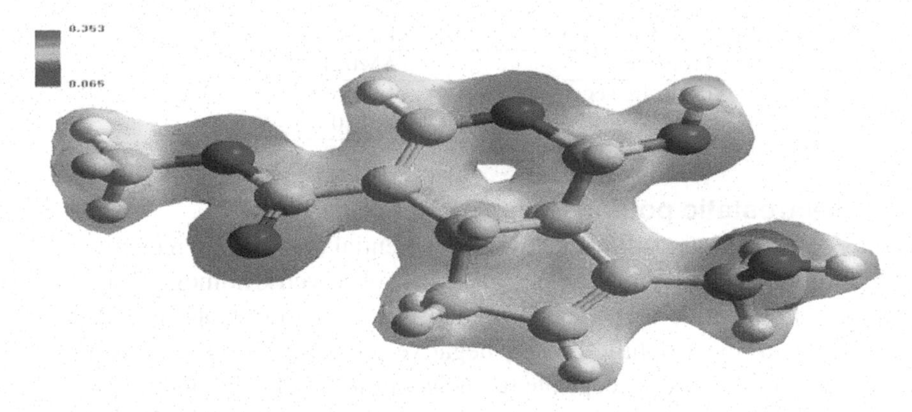

Figure 3.8 Potential electrostatic of genipin: ● carbon, ● oxygen, ● nitrogen, and ○ hydrogen, respectively.

Table 3.4 FTIR of genipin

Assignments		AMBER Frequency (cm^{-1})	Experimental Frequency (cm^{-1}) [25]
O–H	Stretching	3769	3428
C–H	Stretching	3413	2951
H–C–H	Asymmetric stretching	3064	2960–2850
C=O	Stretching	1914	1688
C=C	Stretching	1890	1629
C=C	Stretching (ring)	1863	1440
C–C	Flexion	1516	1500–600
C–O	Flexion	1224	1265

3.3.3 Structural Analysis of Chitosan Cross-Linked with Genipin (C/G)

3.3.3.1 Energy optimization and partition coefficient (Log P)

According to the properties shown in Table 3.5, the cross-linking of chitosan with genipin observed a hydrophilic character due to Log P of –14.10, which agrees with Ashveen et al. which determined the hydrophilicity of the hydrogel chitosan with genipin [26]. The value obtained from the Gibbs free energy of 193.75 Kcal/mol follows that it is a non-spontaneous reaction when compared to individual molecules where an increase in energy can be seen; this is caused by cross-linking, wherein the cross-link covalently chitosan molecule genipin with the amine group change occurs by the amide group, creating new properties and characteristics to the molecule.

Table 3.5 Properties of cross-linking: C/G

Properties	AMBER
Gibbs free energy (Kcal/mol)	193.75
Log P	−14.10

3.3.3.2 Electrostatic potential map

In Figure 3.9, the corresponding electrostatic potential map is shown to cross-link the chitosan with genipin; the electron density observed is homogeneously distributed. The positive areas are attributed to the hydrogen atoms which are susceptible to electrophilic attack; negative areas are those for oxygen atoms, and the amines are susceptible to nucleophilic attack. Given the cross-linking between molecules, a change was observed at partial loads which are modified, due to interbreeding between chitosan and genipin suffering the map of potential changes in the intensities of the area compared to the individual molecules that can be seen in Figures 3.1 and 3.2. In Figure 3.9, you can also observe the formation of the amide, which undergoes rearrangement of the structure of the genipin due to cross-linking.

3.3.3.3 FTIR

Signals of infrared radiation spectrum are shown in Table 3.6, where the presence of stretching vibrations of the OH bond is provided in a range of 3765–3737 cm^{-1}. Moreover, it has a signal from 3418 to 3412 cm^{-1} belonging to the symmetric stretching vibrational mode between the H–N–H. A signal of 1870 cm^{-1} stretching vibration of carbonyl group belonging to

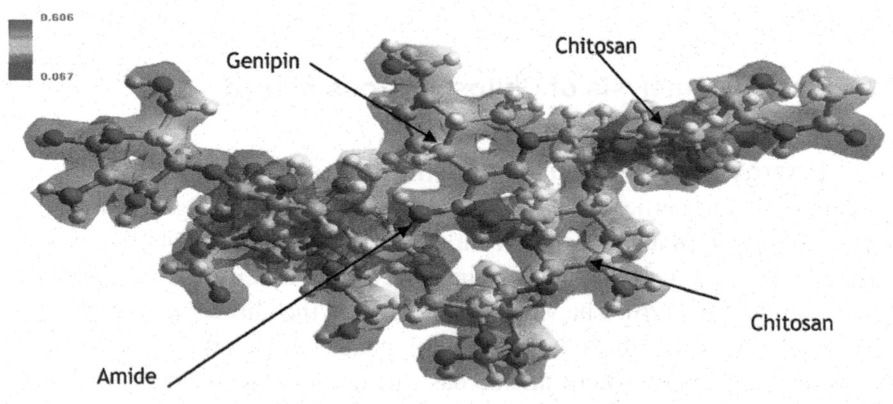

Figure 3.9 Electrostatic potential map of chitosan/genipin: ● carbon, ● oxygen, ● nitrogen, and ○ hydrogen, respectively.

Table 3.6 FTIR of cross-linking of the chitosan/genipin (C/G)

Assignments		AMBER Frequency (cm^{-1})	Experimental Frequency (cm^{-1}) [27]
O–H	Stretching	3765–3737	3650–3200
H–N–H	Symmetric stretching	3418–3412	3500
N–H	Stretching	3284–3277	3500–3300
H–C–H	Symmetric stretching	3199	2960–2850
H–C–H	Asymmetric stretching	3151	2960–2850
C–H	Stretching	2996	4000–2900
C=O	Stretching (chitosan)	1870	1800–1550
C=O	Stretching (genipin)	1853	1800–1550
C=C	Stretching	1793	2000–1500
H–N–H	Flexion	1691–1680	1640–1560
C–C	Flexion	1425	1500–600
C–N	Stretching (cross-linking)	1540	1630
C–N	Stretching	1491	1500–600

the chitosan molecule (C=O) is presented. Similarly, stretching vibration of the carbonyl group (C=O) is shown in a molecule genipin signal of 1853 cm^{-1}. The vibrational mode stretching double bond (C=C) was localized in 1793 cm^{-1}. One of the important bonds is the cross-linking of two molecules of chitosan with a genipin (CN), which belongs to the vibrational mode of stretching a signal of 1540 cm^{-1}, and wherein experimentally is 1630 cm^{-1}. Thus, the results obtained are similar to reported experimentally as shown in Table 3.6 [27].

3.3.4 Structural Analysis of Insulin

3.3.4.1 Energy optimization and partition coefficient (Log P)

In Table 3.7, the Gibbs free energy is appreciated for joining the A and B chains generated greater than those of the individual chains with a value of –228.68 kcal/mol indicating that their formation is spontaneous energy; this is due to disulfide bonds that hold together the chains of insulin. As the negative value

Table 3.7 Properties of the insulin

Properties	Chain: A	Chain: B	Chain: A + B
Gibbs free energy (Kcal/mol)	–63.39	–53.03	–228.68
Log P	–3.69	–2.03	–4.87

of Log P of –4.87, indicating the affinity to polar and according to Muheem solvents et al. It appears that the proteins and peptides are biopolymers of high molecular weight which are commonly hydrophilicity with Log P < 0 [28]. Thus, the Log P value of the insulin molecule is greater than the individual chains because they attach chains increases the number of amino acids, which are linked by peptide bonds.

3.3.4.2 Electrostatic potential map

The electrostatic potential map calculated for the insulin molecule together with the two chains is shown in Figure 3.10. The chains A and B (Figure 3.10) show that the molecules presents a varied electronic distribution, being susceptible to nucleophilic attack areas, where the C=O, OH, and HNH are with the oxygen atoms and more electronegative nitrogen. The difference of

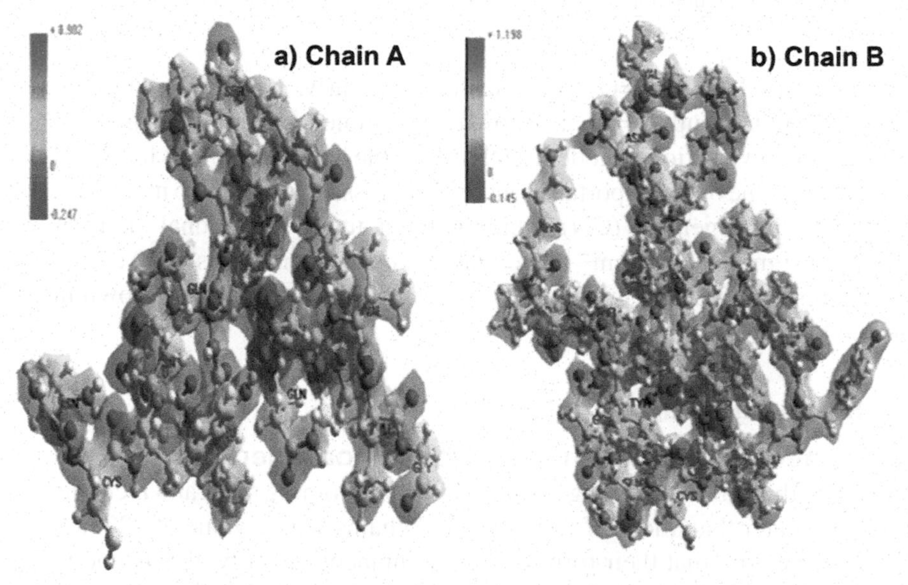

Figure 3.10 Electrostatic potential map of insulin (chains A and B): ● carbon, ● oxygen, ● nitrogen, and ○ hydrogen, respectively.

costs between the chains (A and B) because the chain contains 21 amino acids and the B chain has 30 amino acids, respectively.

In Figure 3.11, the electrostatic potential map for the insulin molecule where A and B chains are linked is shown, and it is observed that most of the molecules are susceptible to nucleophilic attack; this is denoted by the red color and which can be attributed to the similar electronegativity of carbon, oxygen, and nitrogen. Moreover, the difference between the electronic loads of individual and linked chains is due to binding by disulfide bond between these chains.

3.3.4.3 FTIR

For the insulin molecule together the two chains, in Table 3.8 vibrations symmetric stretching between HNH bond signal 3587 cm^{-1}, for a signal of 3236 cm^{-1} are shown stretching vibrations of OH bond is present, likewise it has a signal between 1502 and 1645 cm^{-1} having stretching vibrations of the C=O. A signal of 452 cm^{-1} belonging to stretching vibration of S–S atoms is present. The results were calculated by molecular mechanics with the AMBER method, and these were very similar to the results experimentally shown in Table 3.9 [29].

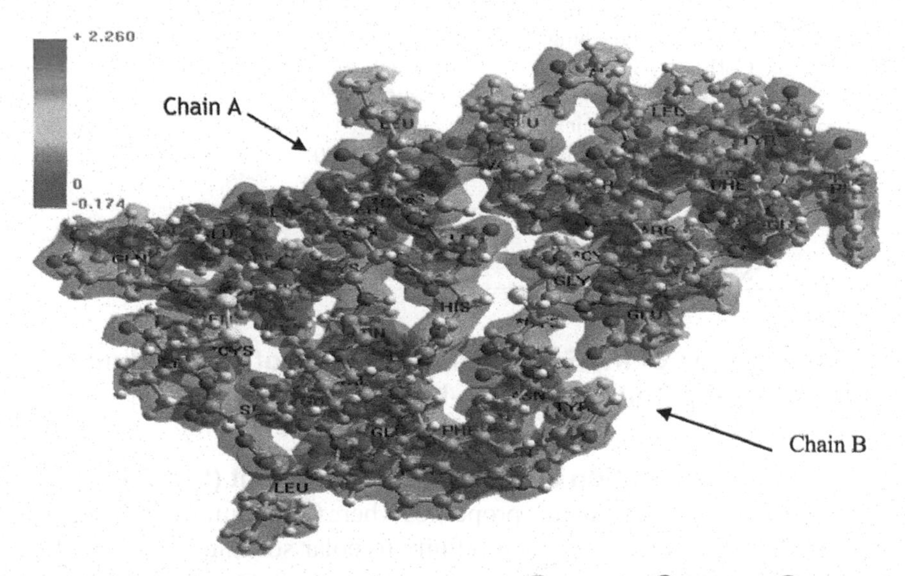

Figure 3.11 Electrostatic potential map of insulin: ● carbon, ● oxygen, ● nitrogen, and ○ hydrogen, respectively.

Table 3.8 FTIR theoretical of insulin

Assignments		AMBER (Chain: A) Frequency (cm^{-1})	AMBER (Chain: B) Frequency (cm^{-1})	AMBER (Chain: A + B) Frequency (cm^{-1})
N–H	Stretching	3442–3849	3724–3936	3710–3871
H–N–H	Estiramiento simétrico	3310	3542	3587
O–H	Stretching	3256	3298	3236
C–H	Stretching	3109	3186	3125
H–C–H	Asymmetric stretching	3072	3141	3125
C=O	Stretching	1563–1617	1538–1836	1502–1645
C–C	Stretching	1195–1371	1132–1416	1366–1416
C–N	Stretching	1034	1096	1011
C–O	Stretching	898	979	850
S–S	Stretching	447	–	452

Table 3.9 FTIR experimental of insulin

Assignments		Experimental Frequency (cm^{-1}) [29]
N–H	Stretching	3500–3300
H–N–H	Symmetric stretching	3400
O–H	Stretching	3650–3200
C–H	Stretching	4000–2900
H–C–H	Asymmetric stretching	2960–2850
C=O	Stretching	1800–1550
C–C	Flexion	962
C–N	Flexion	1098
C–O	Flexion	1267
S–S	Flexion	515

3.3.5 Determination of the Structural Properties of the Binding of Insulin and Chitosan Cross-Linked with Genipin (C/G-insulin)

3.3.5.1 Energy optimization and partition coefficient (Log P)

Table 3.10 shows the thermodynamic properties where the partition coefficient has a value of –40.23, indicating good affinity to polar solvents. The Log P value is greater than insulin and hydrogel as shown in Tables 3.5 and 3.7, respectively; this is due to the absorption of insulin in the hydrogel. It is also noted that the Gibbs free energy obtained was –24.98 Kcal/mol; that is, there is

Table 3.10 Properties of C/G-insulin

Properties	AMBER
Gibbs free energy (Kcal/mol)	–24.98
Log P	–40.23

a spontaneity system and comparing the energy of the cross-linking of chitosan with genipin is lower; this is attributed to the absorption insulin in the hydrogel. Increasing the Gibbs energy tends to increase the partition coefficient (Log P), indicating that there is greater affinity to insulin absorption.

This is attributed to the membranes that have lipid nature, i.e., act as a lipophilic barrier to drug that cannot pass through lipophilic; therefore, if a drug is lipophilic characteristics, it may be absorbed [30].

3.3.5.2 Electrostatic potential map

In the electrostatic potential map calculated for the molecule Q/G-insulin (Figure 3.12), it is seen that most of the two molecules are susceptible to nucleophilic attack; this is denoted by the red color. The attraction between chitosan and insulin is seen through the formation of hydrogen bonds, where the O–H bond of chitosan is attracted by the N–H bond of insulin. Fillers

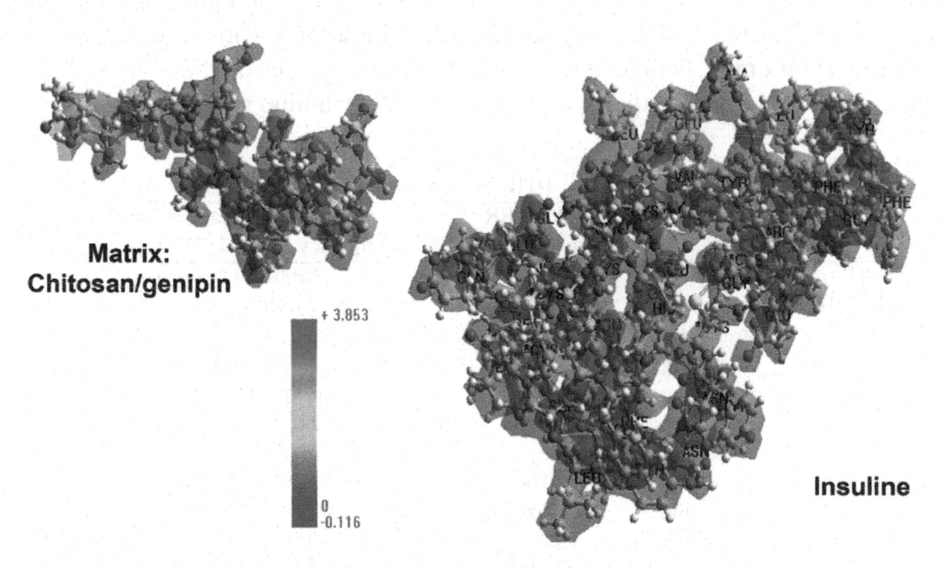

Figure 3.12 Electrostatic potential map of chitosan–genipin/insulin: ● carbon, ● oxygen, ● nitrogen, and ○ hydrogen, respectively.

increase compared to individual molecules as shown in Figures 3.3 and 3.5 due to the hydrogen bond as a bond between hydrogen and an electronegative atom such as oxygen and nitrogen is very polar, so the electron density of the hydrogen is very small. The disulfide bridges (S–S) of insulin make loads change due to the positive charge of the sulfur atom (S) influencing the value of electronic charges.

3.3.5.3 FTIR

The main signals of infrared radiation spectrum are shown in Table 3.11, where the presence of bond stretching vibrations of N–H to a signal of 3488 cm^{-1} is observed. Moreover, it has a signal of 3252 cm^{-1} belonging to the symmetric stretching mode vibrational H–N–H bond.

A signal of 1640 cm^{-1} stretching vibration of carbonyl group belonging to the chitosan molecule (C=O) is presented. The vibrational mode of stretching of C=C bond appeared at 1597 cm^{-1}. The vibrational mode of the C–C, C–N, and C–O bonds is in the area called fingerprint at 1351, 1109, and 938 cm^{-1}, respectively. At the signal of 441 cm^{-1}, stretching vibration of S–S bond occurs. The region between 1800 and 1500 cm^{-1} is the region called amides I and II. In this region, three major bands 1745, 1650, and 1543 cm^{-1}, corresponding to esters, are distinguished. Amide I and amide II, respectively. The band of the peptide bond of the amino acids is between 1700 and 1600 cm^{-1}, which is attibuted to the stretching of the carbonyl group (C=O) [31]. Finally, the electronegativity of the amino group of insulin

Table 3.11 FTIR of C/G–insulin

Assignments		AMBER Frequency (cm^{-1})	Experimental Frequency (cm^{-1}) [31, 32]
N–H	Stretching	3488	3500–3300
H–N–H	Symmetric stretching	3252	3400
O–H	Stretching	3204	3650–3200
C–H	Stretching	2954–3116	4000–2900
H–C–H	Symmetric stretching	2909	2960–2850
C=O	Stretching	1640	1800–1550
C=C	Stretching	1597	2000–1500
C–C	Stretching	1351–1140	1500–600
C–N	Stretching	1109	1500–600
C–O	Stretching	938	1500–600
S–S	Stretching	441	515

with the OH group of chitosan produces the adsorption of insulin in chitosan, generating an attraction hydrogen bond.

3.4 Conclusions

Structural analysis of chitosan, genipin, and the cross-linking of chitosan/genipin, as well as the analysis of the two chains comprising the insulin molecule and chitosan-genipin/insulin, was performed by using the molecular mechanics method AMBER. Likewise, the Gibbs free energy, the partition coefficient (Log P), the electrostatic potential map, and infrared FTIR spectrum were obtained.

1. According to the Gibbs free energy, the non-spontaneous character of the individual molecules of chitosan and genipin that give a positive value was determined, because they do not have a system that is reacting, that is why the AMBER model provides information on positive values. As for the increase in energy, when the chitosan with genipin intertwines is due to the restructuring of the ring to be genipin cross-linking where the amide is formed. However, when this absorption of insulin in the hydrogel generating energy is negative, there is spontaneity for absorption.
2. According to the partition coefficient (Log P), all the molecules are presented with the hydrophilic character (negative values), which indicated that there is affinity between materials to achieve drug absorption into the hydrogel.
3. By analyzing the infrared spectrum, main assignments of joints or functional groups present in the molecules were obtained, whereby the change suffered the molecule when added new atoms or links was established. Chitosan where for major assignments where the presence of stretching vibrations between the OH bond to a signal from 3766 to 3756 cm^{-1} can be seen. These bonds are important because these bonds provide hydrophilic properties that influence the absorption capacity. For genipin, signal of 3769 cm^{-1} corresponded to stretching vibrations of OH bonds; likewise, signals at 1890 cm^{-1} and 1863 cm^{-1} belongs to the vibrational mode of stretching of the C=C rings of genipin. Thus, the appearance of a new bond in the cross-linking (C–N) to a signal of 1540 cm^{-1} belonging to vibrational stretching mode was observed.

In the insulin molecule, the main assignments were located at 3587 cm^{-1} belonging to the vibrational mode–H bond stretching N–H. Similarly, it has bond stretching vibration of S–S where binding is observed between the two chains of insulin to a 452 cm^{-1}. The absorption of insulin in the hydrogel was verified through the O–H bond of chitosan, which was presented to a signal of 3204 cm^{-1}, and the amine (N–H) bond of insulin to a signal from 3488 cm^{-1}.

Some bands were observed shift due to the influence of new atoms, and the calculation method applied (AMBER) should be based on molecular mechanics method.

1. By electrostatic potential map, the electrophilic and nuclophilic areas of the individual molecules were determined; besides, the amide formation was observed, and the adsorption process could determine the attraction of the OH molecule chitosan NH (insulin) bond via a hydrogen bridge formation.
2. Quantum mechanics is not applicable for such systems because the insulin molecule consists of amino acids, which are applicable to molecular mechanics, where the AMBER method is developed for proteins and nucleic acids.

Acknowledgments

Mainly thank God for allowing me to finish my research projects successfully, and thanks to all my family especially my parents for their unconditional support. Special thanks to Dr. Norma Aurea Rangel Vazquez, researcher professor at the Technological Institute of Aguascalientes, who gave me their friendship and trust for the realization of this project. I also thank the Council for Science and Technology (CONACYT) for their support for this project and the Technological Institute of Aguascalientes especially researchers and teachers of the Department of Graduate Master of Science in Chemical Engineering for his contributions to enrich the research of this project.

References

[1] Muñoz, R. G. (2012). Obtención de hidrogeles de quitosano a partir del micelio de aspergillus niger Para la liberación controlada de fármacos. Tesis de Maestría en Ciencias-Químicas, Universidad del Valle, Colombia.

[2] Kumar, A., et al. (2007). Smart polymers: physical form & bioengineering applications, *Progress Poly. Sci.* 32, 1205.

[3] Okuyama, K., Noguchi, K., Kanenari, M., Egawa, T., and Osawa, K. (2000). Structural diversities of chitosan and its complexes. *Carbohydrate Polym.* 41, 237–247.

[4] Martin, P. G. (1995). Applications and environmental aspects of chitin and chitosan. *Pure Appl. Chem. J. Macromol. Sci., Part A: Pure Appl. Chem.* 32, 629–640.

[5] Acosta, N., Jiménez, C., Borau, V., and Heras, A. (1993). Extraction and characterization of chitin from crustaceans. *Biomass Bioenergy* 5, 145–153.

[6] Zeng, D., Luo, X., and Tu, R. (2012). Application of bioactive coatings based on chitosan for soybean seed protection. *Intl. J. Carbohydrate Chem.* 2012, 5.

[7] http://www.eis.uva.es/~macromol/curso05-06/medicina/hidrogeles.htm/ Consultado Noviembre 2014

[8] Sánchez, A., Sibaja, M., and Vega, J. (2007). Síntesis y caracterización de hidrogeles de quitosano obtenido a partir de camarón langostino (pleuroncodesplanipes) con potenciales aplicaciones biomédicas. *RevistaIberoamericana de polímeros* 8, 241–267.

[9] Bulter, M. F., Ng, Y. F., and Pudney, P. D. (2003). Mechanism and kinetics of the crosslinking reaction between biopolymers containing primary amine groups and genipin. *Polym. Sci.* 41, 3941–3953.

[10] Reyes, O. F., Rodriguez, M. G., Aguilar, M. R. and Garcia, S. J. (2012). Comportamiento reológico de geles biodegradables para aplicaciones en medicina regenerativa. *Biomecanica* 20, 7–19.

[11] Harris R. E. (2010). Quitosano un Biopolímero con Aplicaciones en Sistemas de Liberación de Fármacos. Universidad Complutense de Madrid, Tesis de Doctorado Madrid.

[12] Arredondo, P. A., and Londono, L. M. (2009). Hidrogeles. Potenciales biomateriales para la liberación controlada de medicamentos. *RevistaIngenieríaBiomédica*, 3, 83–94.

[13] Mengatto, L. N. (2010). Administración de Fármacos por Vía Transdérmica. Tesis de Doctorado en Ciencias Biológicas, Universidad Autónoma de Litoral, Argentina.

[14] Poland, C. A., Read, S. A., Varet, J., Carse, G., Christensen, F. M. (2013). Dermal absorption of nanomaterials. Danish Ministry of the Environment, Environmental Project No. 1504.

[15] http://www.idf.org/diabetesatlas/5e/es/que-es-la-diabetes?language=es/ Consultado Enero 2015

[16] http://www.who.int/mediacentre/factsheets/fs312/es/ Consultado Enero 2015
[17] Instituto Nacional de Estadística y Geografía (INEGI). Mujeres y Hombres en México 2014.
[18] Olivares, R. J., Arellano, P. A. (2008). Bases Moleculares de las Acciones de la Insulina. *REB*. 27, 9–18.
[19] Lewars E. (2011). *Computational chemistry. Introduction to the theory and applications of molecular and quantum mechanics*, 2nd Edn. (Berlin: Springer).
[20] Olivares, R. J., and Arellano P. A. (2008). Bases Moleculares de las Acciones de la Insulina. *REB*. 27, 9–18.
[21] Trung, T., Wah Thein-Ha, and Qui, N. (2006). Functional characteristics of shrimp chitosan and its membranes as affected by the degree of deacetylation. *Bioresource Technol*. 9, 659–663.
[22] Lárez, V. C. (2003). Algunos usos del quitosano en sistemas acuosos. *Revista Iberoamericana de Polimeros*, 4, 91–109.
[23] Balanta, D., Grande, D., Zuluga, F. (2010). Extracción, identificación y caracterización de quitosano del micelio de aspergillus niger y sus derivados aplicaciones como material bioadsorbente en el tratamiento de aguas. *Revista Iberoamericana de Polímeros*, 11, 297–316.
[24] Yoo, J. S., Kim, Y. K., Kim, S. H. (2011). Study on Genipin: A new alternative natural crosslinking agent for fixing heterograft tissue. *Korean J. Thorac. Cardiovasc. Surg*. 44, 197–207.
[25] Álvarez, O. G. (2013). Extracción, caracterización y valoración de genipina a partir del fruto de la genipina americana. Tesis de Licenciatura de Químico Farmacéutico. Universidad ICESI, Facultad de Ciencias Naturales, Departamento de Ciencias Químicas Santiago de Cali.
[26] Sadashiv, A. M., Jen, L. W., and Suresh, P. S. (2015). Application of polymeric nanoparticles and micelles in insulin oral delivery. *J. Food Drug Anal*. 23, 351–358.
[27] Santoni, N., Matos, M., and Muller C. (2008). Caracterización de hidrogeles de quitosano entrecruzados covalentemente con genipina. *RevistaIberoamericana de Polímeros* 9, 326–330.
[28] Muheem, A., Shakeel, F., Asadullah, M. et al. (2016). A review on the strategies for oral delivery of proteins and peptides and their clinical perspectives. *Saudi Pharmaceutical J*. 24, 413–428.
[29] Tah, B., Pal, P., and Roy, S. (2014). Quantum-mechanical DFT calculation supported Raman spectroscopic study of some amino acids

in bovine insulin. *Molecular and Biomolecular Spectroscopy*. 129, 345–351.

[30] http://www.elergonomista.com/galenica/absorcion.htm/ Consultado Enero 2016

[31] Kikot, G. E. (2012). Caracterización bioquímica, fenotípica y molecular de aislamientos de Fusarium graminearum provenientes de la región pampeana en relación a la patogenicidad. Tesis de Doctorado, Universidad Nacional de la Plata.

[32] http://depa.fquim.unam.mx/amyd/archivero/InfrarrojoTablas_31544.pdf/ Consultado Agosto 2015

4

Analysis and Molecular Characterization of Organic Materials for Application in Solar Cells

Norma-Aurea Rangel-Vázquez[1]
and Ediht-Sofia Martínez-Rodríguez[2]

[1]PCC, Aguascalientes, Mexico
[2]Chemical Engineering Department, Master in Science in Chemical Eng. Technological Institute of Aguascalientes, Mexico

Abstract

Synthesizing new materials that resist degradation by environmental agents can be costly in time and money at laboratory level, but with the molecular modeling, which includes theoretical computational techniques and methods for modeling, mimic and predict the behavior of molecules that can perform experiments with a considerable saving of resources. Applications of nanotechnology require a thorough knowledge of the theoretical and computational aspects of all kinds of materials at the nanoscale; molecular modeling by describing systems at the atomic level is what allows us to obtain that knowledge. So molecular modeling can be created by analyzing and characterizing nanocomposite organic materials and combining them with those that are resistant to degradation, to observe their behavior in environmental conditions to which they are exposed by their application in organic solar cells.

The polymers are good matrices for nanocomposites, and together with the load, combine their different attributes for a material with improved properties. They were used in this project: the polymer poli[4,4-bis(2-etilhexil)ciclopenta[2,1-b;3,4-b′]ditiofeno-2,6-diyl-alt-2,1,3-benzotiadizol-4,7-diyl] (PCPDTBT), because it represents one of the best π-conjugated polymers, electron donor, with response spectrum which extends to 900 nm.

The fullerene is an allotrope of carbon and has shown excellent properties as an electron acceptor.

Polyethylene is a chemical-resistant thermoplastic, to prevent deterioration with UV light are added protective and being a non-hygroscopic polymer humidity not significantly affect you. The aforementioned materials can be arranged through orders, and PCPDTBT and fullerenes have moderate costs relative to other polymers with similar uses, and polyethylene is the cheapest and most versatile polymers because you may be adapted depending on the use made required.

Keywords: PCPDTBT, fullerenes, molecular simulation.

4.1 Introduction

4.1.1 Computational Chemistry

Computational chemistry is a science that links theoretical methods (physical and mathematical) and computational techniques, with the aim of representing chemical systems at the atomic level and calculate various properties that allow us to understand and predict the behavior of such systems to different conditions. Over the years, due to increased technology, computational chemistry has been increasing its importance due to the considerable savings in economic, human, and time resources; also, for this reason, they performed the calculations which are more accurate each time comparable with those obtained experimentally [1–3].

Computational chemistry can be divided into two study areas: molecular mechanics and quantum mechanics, which are distinguished by the considerations and the equations used to perform various calculations. For molecular mechanics, the system is analyzed based on classical mechanics and perform calculations only on the position of the nuclei taken into account, and it is assumed that the electrons are distributed optimally. On the other hand, quantum mechanics is based on the Schrödinger's equation to obtain data about the system and therefore takes into account each electron of the same [4, 5].

4.1.1.1 Molecular mechanics (MM)

The molecular mechanics (MM) applies the laws of classical mechanics for energy, vibrational spectra, enthalpies of formation, and structures of molecules found in basal state. Other names used for molecular mechanics are force fields and fields of potential, which refer to the equations and parameters used to describe the behavior of the system [5, 6].

To perform calculations, MM considers the atoms as spheres of certain mass, where the position of the nuclei is most important, without explicitly considering the electrons. The MM ranks atoms using a strategy called "types of atoms", thus intended that the calculations for atoms can be extended to other systems where they themselves are; for the MM behavior of each atom it depends on the environment that surrounds it; thus, creating types of atoms is an attempt to explain this behavior [7–9].

If deemed atoms as masses and bonds as springs, then Hooke's law is used ("deformation experienced by a body is directly proportional to the stress produced") [10] to calculate the force constants; therefore, the different bonds have several constants. In Hooke's law (Equation 4.1), f is the force required to stretch the spring, x elongation, and k the force constant, so the potential energy of the spring is described as follows [11].

$$f = kx^2 \tag{4.1}$$

From the point masses and springs, the MM describes the total energy of a molecule as a sum of energies from different contributions: type and bond angles, torsion angles, van der Waals force, electrostatic repulsions, and the associations by hydrogen bonds (Equation 4.20) [8, 9].

$$E_{\text{total}} = E_{\text{stretching}} + E_{\text{flexion}} + E_{\text{torsion}} + E_{\text{VdW}} + E_{\text{elec.}} + E_{\text{H}} \tag{4.2}$$

The potential energy of stretching bond between two atoms is that required to stretch or compress the bond is shown in Figure 4.1 and can be expressed mathematically according to Hooke's law (Equation 4.3), where r_{eq} is the equilibrium length, r is the distance between the two atoms, and k_{e} is stretching force constant [12].

$$E_{\text{stretching}} = \frac{1}{2}k_{\text{e}}(r - r_{\text{eq}})^2 \tag{4.3}$$

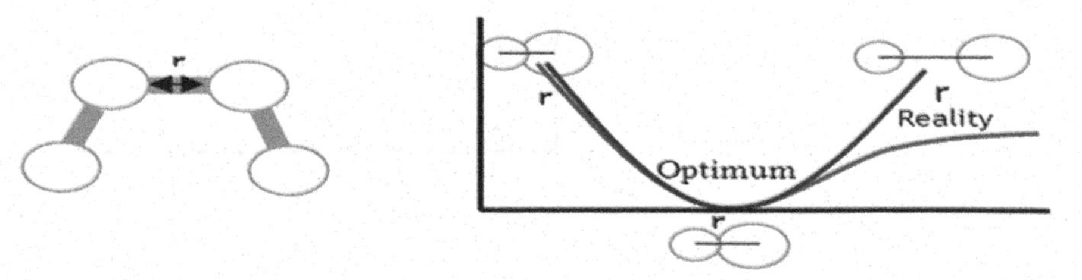

Figure 4.1 Energy potential stretch.

Figure 4.2 and Equation 4.4 show the flexion potential energy, i.e., the energy required to bend a bond from the balance; the mathematical treatment similar to that bond, that is, is used the Hooke's law again [8, 11].

$$E_{\text{flexion}} = \frac{1}{2} k_{\text{f}} \left(\theta - \theta_{\text{eq}}\right)^2 \tag{4.4}$$

It is noteworthy that, in trying to capture the movement of an atom between two planes, the energy of torsion is obtained; this is the energy it takes to turn around the bonds; this is important to single bonds because when there are double or triple bonds, these are very rigid and not allow movement as shown in Figure 4.3.

To express the energy of torsion mathematically, Fourier series are used, keeping the structure of Hooke's law (Equation 4.5) [8, 11].

$$E_{\text{torsion}} = \frac{1}{2} k_{\text{t}_1} \left(1 + \cos\omega\right)^2 + \frac{1}{2} k_{\text{t}_2} (1 - 2\cos\omega)^2 + \frac{1}{2} k_{\text{t}_3} (1 + 3\cos\omega)^2 \tag{4.5}$$

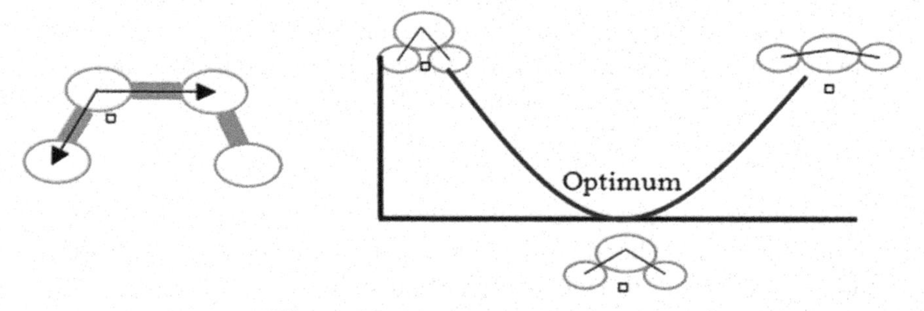

Figure 4.2 Flexion potential energy.

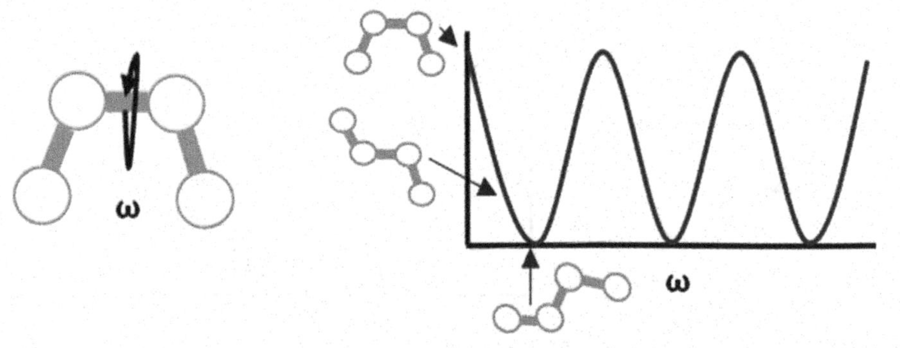

Figure 4.3 Energy of torsion.

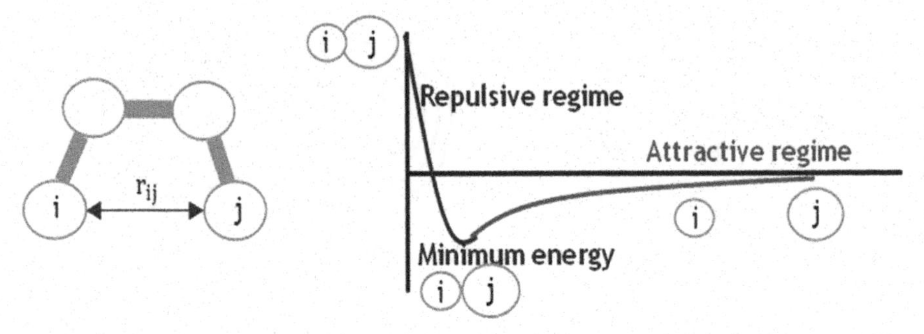

Figure 4.4 Van der Waals energy.

The potential energy of van der Waals (VDW) occurs between atoms that are not bonded by the electron clouds; these interactions are those that give shape to the molecule, if an atom is far from other interactions tend to scratch, a medium-range attractive forces dominate and if they are very close, are strongly dominant repulsive, so the latter are those that primarily provide the shape the molecule, as shown in Figure 4.4 [13].

VDW energy normally calculated for atoms "connected" with at least four atoms, and for reducing the computational costs, the equation Lennard-Jones potential (Equation 4.6) is used, where ε is the minimum energy, r_0 is the sum of the VDW radii, and r_{ij} is the distance separating the two nuclei in a; the first part in parenthesis corresponds to the repulsive forces and the second to attractive forces [11].

$$E_{\mathrm{VdW}} = \varepsilon \left[\left(\frac{r_0}{r_{\mathrm{ij}}} \right)^{12} - \left(\frac{r_0}{r_{\mathrm{ij}}} \right)^{6} \right] \tag{4.6}$$

Energy from the electrostatic term is described by Coulomb interactions between two atoms, as in the case of VDW energy; if atoms are close, then the energy is repulsive. If the bonds of a molecule are polar, then there are partial electrostatic charges on atoms and can be represented mathematically by a function of Coulomb potential (Equation 4.7), where Q_{i} and Q_{j} are the atomic partial loads, *D* is the dielectric constant, and r_{ij} is the distance separating the two nuclei in Å, as shown in Figure 4.5 [8, 11].

$$E_{\mathrm{elec.}} = \frac{Q_{\mathrm{i}}\, Q_{\mathrm{j}}}{D\, r_{\mathrm{ij}}} \tag{4.7}$$

Equation 4.7 is also advisable to understand the electrostatic attraction between ions. For molecules with permanent dipole moments, it is appropriate

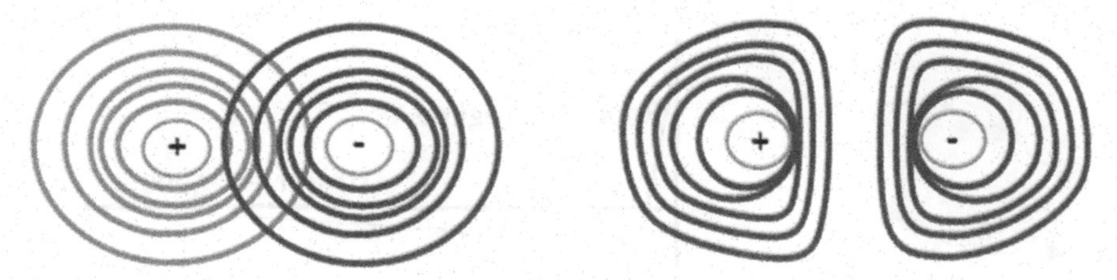

Figure 4.5 Energy of Van der Waals.

to include dipoledipole interactions using the equation Jeans as seen in Equation 4.8, where μ_i and μ_j are dipoles, the angles that each dipole makes with the line joining the atoms *j* are $\alpha_i\alpha_j$, and φ is the angle between the dipoles [8, 12].

$$E_{elec.} = \frac{\mu_i\, \mu_j}{D\, {r_{ij}}^3}\ (\cos\ \varphi - 3\cos\ \alpha_i\alpha_j) \tag{4.8}$$

With regard to energy contributions by hydrogen bonds (H), are generated due to the attraction inter- and intra-molecular which always involves a hydrogen, so that energy from bridge H is very weak, thus generating a dipolar character. This energy is expressed mathematically using Lennard-Jones equation (Equation 4.9) [13, 14].

$$E_H = \varepsilon \left[\left(\frac{r_0}{r_{ij}}\right)^{12} - \left(\frac{r_0}{r_{ij}}\right)^{10} \right] \tag{4.9}$$

Furthermore, the process by which the optimum values of the parameters are determined is called parameterization. This is accomplished by adjusting the parameters to reproduce the behavior of molecules. These parameters are obtained from either experimental data or data obtained from *Ab-initio* methods [15]. The parameterization is one of the key points and requires a careful choice of the experimental data because the success and reliability of a force field depend on making a good parameterization constants and types of atoms.

4.1.1.1.1 *AMBER model*

Assisted model building with energy refinement (AMBER) is a set of force fields designed and configured especially for the simulation of proteins and nucleic acids, and was developed at the University of California by Peter Kollman and his research group; this model became the force field most used academically, so the "descendants" of Peter Kollman and equipment and other researchers have done work on improving codes, and parameter settings remain the most widely used at this time: the FF94, FF96, and FF99. As a method of molecular mechanics, the mathematical structure corresponds to a sum of energies which provides the end of the calculations and the total energy of the molecule. The mathematical form of AMBER is expressed in Equation 4.10 as follows [16–19]:

$$\begin{aligned} E_{\mathrm{Total}} = & \sum_{\text{stretching}} K_{\mathrm{e}}(r - r_{\mathrm{eq}})^2 + \sum_{\text{flexion}} K_{\mathrm{f}}(\theta - \theta_{\mathrm{eq}})^2 \\ & + \sum_{\text{dihedral}} \frac{V_{\mathrm{n}}}{2}[1 + \cos(n\Phi - \Phi_0) + \sum_{\text{VdW}} \left[\frac{A_{\mathrm{ij}}}{R_{\mathrm{ij}}^{12}} - \frac{B_{\mathrm{ij}}}{R_{\mathrm{ij}}^{6}}\right] \\ & + \sum_{\text{electrostatic}} \frac{q_{\mathrm{i}} q_{\mathrm{j}}}{\varepsilon R_{\mathrm{ij}}} + \sum_{\text{H bond}} \frac{C_{\mathrm{ij}}}{R_{\mathrm{ij}}^{12}} - \frac{D_{\mathrm{ij}}}{R_{\mathrm{ij}}^{10}} \end{aligned} \tag{4.10}$$

where K is the force constant of each energy, V_{n} is the highest energy conformation-cis, φ is the dihedral angle, φ_0 is the phase angle, A_{ij} and B_{ij} are parameters for the interaction of pairs of atoms, and R are the radii of balance; the terms are similar to those seen in the previous section of molecular mechanics [19, 20].

Constants and other terms are parameterized as V_{n}, A_{ij}, and B_{ij}, while these are calculated as follows: $A_{\mathrm{ij}} = \left(\frac{r_{\mathrm{i}}^*}{2} + \frac{r_{\mathrm{j}}^*}{2}\right)^{12} \sqrt{\varepsilon_{\mathrm{i}}\, \varepsilon_{\mathrm{j}}}$, and $B_{\mathrm{ij}} = 2\left(\frac{r_{\mathrm{i}}^*}{2} + \frac{r_{\mathrm{j}}^*}{2}\right)^{6} \sqrt{\varepsilon_{\mathrm{i}}\, \varepsilon_{\mathrm{j}}}$, where r_{i}^* is the minimum energy of separation for two atoms of a certain type and ε is the well depth, both are parameters which are also found in AMBER.

Assisted model building with energy refinement, energy functions for stretching and bending, is only quadratic. In the energy function by electrostatic force scale, the dielectric is usually 1 and standard loads can be placed manually if so required. For the function of hydrogen bonds, AMBER replaces only exponents to 10 or 12 in the van der Waals equation; the coefficients C_{ij} and D_{ij} are obtained of parameters. Finally, for the lone pairs, solitary AMBER adds to the sulfur atom pairs and calculates the interactions as if they were a specific type of atom [21].

The parameter values for protein systems were optimized for H, C, N, O, and S; the value of P is not optimized at all for these systems; on the other hand, for nucleic acids, default settings were optimized for H, C, N, O, and P. With the right choice of parameters, the calculations at the quantum level on many conformations of the structures must match those calculated AMBER. Thus, this simulation model has evolved over time as a technological tool that enables researchers to model quickly and accurately large molecules such as biomolecules [17, 21].

4.1.1.2 Quantum mechanics

Quantum mechanics (MC) is the science of the very small, studies the microscopic scale phenomena, and explains the behavior of the energy and matter at the atomic level or particles. It was developed by Erwin Schrödinger, Werner Heisenberg, Paul Dirac, and other scientists in the 1920s [22]. The most important concepts of the MC are as follows:

- The energy is not exchanged continuously, but there is a minimum amount involved (energy quantization).
- You cannot be set the position and velocity of a particle (uncertainty principle). You can only have a probability that the particle is in a place in a moment. As a wavefunction of which is theoretically extract all magnitudes of motion necessary it is used [23].

In Equation 4.11, the three-dimensional Schrödinger equation is deduced that obeys the wavefunction associated with a particle.

$$ih\,\frac{\partial\Psi\;(x,t)}{\partial t} = -\frac{h^2}{k}\,\nabla^2\;\Psi\;(\mathbf{r},\mathrm{t}) + V\;\;(\mathbf{r},t)\;\;\Psi\;(\mathbf{r},\mathrm{t}) \tag{4.11}$$

By its nature, the wavefunction Ψ (*x*, *t*) is complex and therefore cannot be measured with real instrument. For a given problem, you have to provide the shape of the potential [*V* (*x*), if analyzed on the dimension *x*], and this determines the particular form of the differential equation satisfied by the wavefunction.

Actually, Ψ (*x*, *t*) is nothing more than a calculation tool that only has meaning in the context of the Schrödinger's theory; this does not mean that the function lacks physical interest because the wavefunction contains all information on the particle study and is compatible with the principle of uncertainty. For that information, Max Born proposed a relationship where the wavefunction squared [|Ψ (*x*, *t*) |^2] gives the probability of finding the particle at point *x* at time *t*: $P(x,t)dx = \frac{|\Psi(x,t)|^2 dx}{\int_{\text{todo } x} |\Psi(x,t)|^2 dx}$ [24].

If the particle is an electron which indicates where you can find an electron, this is referred to as orbital; by solving the Schrödinger's equation, you can get a set of molecular orbitals and the energy of each of them (the most important are HOMO and LUMO) plus the total energy of the molecule, the electron density, the loads on each atom, electrostatic potential spin densities, among others [21, 24]. Quantum mechanics is divided into *Ab-initio* methods ("from the beginning") and semiempirical methods.

The *Ab-initio* methods are rigorous methods that solve the Schrödinger's equation without the help of experimental information but only using any corrections called correlations; semiempirical method is described in the next section [4].

4.1.1.3 Semiempirical methods

The semiempirical methods (MS) can analyze large molecules relatively fast and produce accurate results when applied similar to those used for parameterization systems; this is achieved due to the omission or parameterization of integrated certain and replacing them with experimental data; for example, modern methods are based on the neglect of diatomic differential overlap (NDDO). This method ignores the array of overlapping and replaces it with a unitary matrix; this can replace the secular equation of Hartree-Fock (4.1) by Equation 4.12, where H is the secular determinant, S is the matrix of overlap, and E is the set of eigenvalues [25, 26].

$$[H - ES] = 0\ (Ec.1) \rightarrow [H - E] = 0 \tag{4.12}$$

These approaches will greatly simplify the calculations in systems' study. All methods are more or less the same structure: They are based on the Hartree-Fock (which seeks solutions to the Schrödinger's equation) and are methods of self-consistent field (SCF). The strategy is simple: the MS starts with formalisms *Ab-initio* or rudimentary, after quite drastic assumptions introduced to accelerate calculations and to compensate empirical parameters obtained from experimental data or reliable theoretical incorporated. Standard equations are simplified by approximate integrals whose purpose is to omit the whole three-center and four-center two electrons [27].

Subsequently, the integrals do not cancel and associated parameters introduced are evaluated; selection of appropriate methods of parametric expressions is guided by the integral analytical analysis or intuition; this serves to determine the values of the optimum parameters by calibration with the reference data. Therefore, the quality of the results depends on the care and effort made in the parameterization.

In practice, the MS may be useful and easily performed for 1,000 molecules containing atoms other than hydrogen. For larger molecules, it is recommended to use alternative approaches such as linear scaling methods and hybrid methods of molecular mechanics and quantum mechanics [28].

4.1.1.3.1 *Parametric method 3*

Parametric method 3 (PM3) is a semiempirical method resulted from the combination of theoretical framework and optimization of parameters MNDO and AM1, hence its name considering the above methods as 1 and 2. In the initial work assembly, parameters consisted of 18 parameters for each element than hydrogen (for which are only 11) [29].

The purpose of the parameters is to accurately represent the values that are experimentally observed by chemical properties, which are obtained using as reference functions dimensionless quantities, and represent the experimentally observed phenomena. Most reference functions are derived from experimental data, but sometimes results of high-level *Ab-initio* are also used along with experimental data, particularly for geometries. Property sets are as follows: heat of reaction and formation, dipole moments, ionization potentials, and molecular geometries. By using a lot of reference functions, errors can be minimized [30].

Under the semiempirical approach, the standard equations SCF-MO of Hartree-Fock are simplified by approximate integrals, which are designed to bypass all integrals three- and four-center two electrons. Assuming $\mu^A \nu^B d\tau = \delta_{AB}\, \mu^A\, \nu^B\, d\tau$, where μ and ν are atomic orbitals of an atom A or B and $d\tau$ is the volume element. This keeps the integrals of uncenter and a large number of comprehensive two centers. Molecular orbital (MO) ψ_i is expressed as a combination of atomic orbitals (OAs, φ_i) [27].

Expansion coefficient $C_{\mu i}$ and orbital energy ε_i are obtained from the solution of the secular equations: $\psi_i = \sum_\mu c_{\mu i}\phi_\mu$ y $0 = \sum_\nu (F_{\mu\nu} - \delta_{\mu\nu}\, \varepsilon_i)\, c_{\nu i}$. Following the comprehensive approach, the elements of the matrix of Fock ($F_{\mu\nu}$) contain only terms of uncenter and two centers, which defined as follows: $F_{\mu^A \nu^A} = H_{\mu^A \nu^A} + \sum_{\lambda^A} \sum_{\sigma^A} P_{\lambda^A \sigma^A} [(\mu^A \nu^A, \lambda^A \sigma^A) - \frac{1}{2}(\mu^A \nu^A, \lambda^A \sigma^A)] + \sum_B \sum_{\lambda^B} \sum_{\sigma^B} P_{\lambda^B \sigma^B} (\mu^A \nu^A, \lambda^B \sigma^B)$ y $F_{\mu^A \nu^A} = H_{\mu^A \nu^A} - \frac{1}{2} \sum_{\lambda^A} \sum_{\sigma^B} P_{\lambda^A \sigma^B}\, (\mu^A \nu^A,\ \lambda^B \sigma^B)$, where $H_{\mu\nu}$ are the elements of a-electron Hamiltonian, $P_{\lambda\sigma}$ is the density matrix, and $(\mu\nu,\ \lambda\sigma)$ denote an integral two electrons. The total energy (E_{tot}) of a molecule is the sum of the electronic energy (E_{el}) and repulsions core–core ($E_{\mathrm{AB}}^{\mathrm{core}}$), which is composed of the term electrostatic ($E_{\mathrm{AB}}^{\mathrm{coul}}$) and an additional effective term ($E_{\mathrm{AB}}^{\mathrm{eff}}$), such that $E_{\mathrm{el}} = \frac{1}{2} \sum_\mu \sum_\nu P_{\mu\nu}\, (H_{\mu\nu} + F_{\mu\nu})$ and then $E_{\mathrm{tot}} = E_{\mathrm{el}} + \sum_{A<} \sum_B E_{AB}^{\mathrm{core}}$ [27, 30].

The effective term of the potential of a pair of atoms E^{eff}_{AB} tries to take into account the Pauli repulsion exchange and also compensate for the errors introduced by other assumptions. Finally, resonance integrals $\beta_{\mu\nu} = H_{\mu^A\nu^B}$ are taken as the corresponding overlap integrals. This term PM3 is represented by a more flexible function with several adjustable parameters [27].

What makes it different and better to PM3 than its two predecessors is that it takes the whole repulsive electron uncenter as parameters to be optimized, the parameters for each element have two Gaussian, it uses a method for automatic parameterization, and also, the parameters were derived by comparing a very large and varied amount of experimental data with calculated molecular properties. The elements for which PM3 is parameterized are shaded in Figure 4.6 [21, 25].

4.1.2 Composites

Composite materials (also called composites) are those which are formed by reinforced with particles of various material particles, being the main feature that the resulting material has improved properties regarding the individual materials. The particles are added to improve properties such as elastic modulus of the matrix, fire behavior, heat resistance, and yield strength, among others. To scale the particle size to the nanometer level, it has been shown that new properties can be obtained in the materials. Composites are formed by two parts: the matrix and reinforcement (or load); the first is the continuous phase and the second is the dispersed phase [31, 32].

The main functions of the matrix are as follows: provide structural stability to the material by transmitting loads to the reinforcement, protect the reinforcement of chemical and mechanical deterioration, and define the physical and chemical properties. The reinforcement instead is generally used to increase the mechanical strength and rigidity, but can also be used to improve high-temperature performance and abrasion resistance [33, 34].

The composites can be classified according to the matrix: ceramic (CMC), metal (MMC), and polymer (PMC). CMC is used when the material will be exposed to very high temperatures, the MMC has its greatest range of applications in the automotive area, and finally, the PMC is the most used because the polymers form excellent host matrices.

Moreover, the reinforcement may be in the form of particles or fibers and between smaller particles and are homogeneously distributed in the matrix which is more effective than the material. Also, the success of composite depends on the interface because the better sticking together the atoms

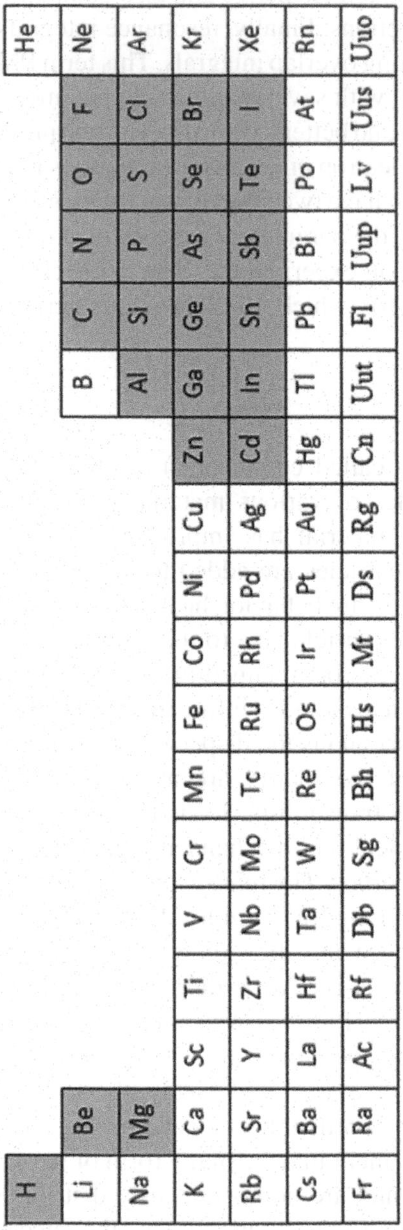

H																	He
Li	Be											B	C	N	O	F	Ne
Na	Mg											Al	Si	P	S	Cl	Ar
K	Ca	Sc	Ti	V	Cr	Mn	Fe	Co	Ni	Cu	Zn	Ga	Ge	As	Se	Br	Kr
Rb	Sr	Y	Zr	Nb	Mo	Tc	Ru	Rh	Pd	Ag	Cd	In	Sn	Sb	Te	I	Xe
Cs	Ba	La	Hf	Ta	W	Re	Os	Ir	Pt	Au	Hg	Tl	Pb	Bi	Po	At	Rn
Fr	Ra	Ac	Rf	Db	Sg	Bh	Hs	Mt	Ds	Rg	Cn	Uut	Fl	Uup	Lv	Uus	Uuo

Figure 4.6 Elements for PM3 parameterized model.

at each part has improved strength properties, and otherwise, there may be deterioration of mechanical properties [35–37].

Nanocomposites are composites in which at least the size of one of the phases is in the nanometer range ($1\mathrm{nm} = 10 - 9\mathrm{m}$). A nanocomposite also divided into matrix and load, the load being located in the nanometric size [38]. Despite being a relatively new technology, applications of nanocomposites are numerous, including the generation of new materials such as improving the performance of known devices. Some of these applications are as follows: high-performance catalysts, technological data storage, optical fibers, chemical sensors, magnetic devices, photo-electrochemical applications, conversion of light energy, aerospace and aeronautical materials, and many more [39, 40].

4.1.2.1 Polymer matrix

The polymers form excellent host matrices for the manufacture of composites, due to the fact that they are easily adapted to produce multiple physical properties. Selecting a polymeric matrix is usual and primarily through mechanical, thermal, electrical, optical, and magnetic behavior. Moreover, other properties such as the balance hydrophobicity/hydrophilicity, chemical stability, biocompatibility, chemical and optoelectronic properties, and chemical functions are considered. The polymers can also allow easier and better conformation processing of composite materials [41].

The properties of the polymeric matrices (PM) depend on the matrix, the reinforcement, and the interface; this makes for manufacturing not only the type of matrix and reinforcement, but also proportions, the geometry of the reinforcement, and the nature of the interface, so each point must be carefully calculated to obtain a PMC optimized for the application that you want to give [42, 43].

The PM, sometimes also called organic matrices, can be classified according to whether or not cross-links are presented, therefore, fall into, thermoset and thermoplastic. Thermosetting is easier to process, but its shrinkage upon curing is also high; as this reaction is exothermic, it can make damage to the material, and it has good chemical and corrosion resistance, but the mechanical properties are medium to low, and although its toughness is high, it is more fragile than the thermoplastic. While, thermoplastic, are more difficult to process due to their high molecular weight also it provides excellent mechanical properties, good chemical resistance, low moisture absorption, can be repaired to soften with heat the affected part and also can be recycled [35].

Today because of the latter feature, the PM and thermoplastic are replacing the thermoset. Processing techniques for PM–thermoset are special designs, while processing techniques for PM–thermoplastic are those under which the matrix is processed individually, i.e., injection molding, extrusion, rotomolding, or compression. Typically, more loads are spiked to PM particles: spherical (silica, metal, and other organic and inorganic particles), fibers (nanofibers and nanotubes), and lamellar (carbon, graphite, silicates, aluminosilicates laminates, and other sheet goods) [44].

In general, the load will provide high strength and rigidity to the polymer matrix. By incorporating conductive fillers in the PM as a second phase, electrical conductivity is achieved in the resulting composite materials. The interface of PM is a length where the load is transferred between the matrix and reinforcement. Its extension is variable and depends on the design, allowing bonding force range from very large to weak friction forces [33, 44].

First, it causes the PMC more rigid but fragile; if it is a weak bond, stiffness decreases but increases the toughness, and in the most extreme case of weakness, components are separated and the material is lost their properties. The most desirable and advisable is to find a compromise between the limits of strength and weakness. At present, the PM to the automotive, marine, and construction, biomedicine, the electric field (for conductive and nonconductive) extend, among others [35, 45].

4.1.3 Polymers

4.1.3.1 High-density polyethylene

The characteristics of polyethylene depend on the arrangement of their molecular chains; also, the number, size, and type of branching or side chains largely determine the properties of density, stiffness, tensile strength, flexibility, hardness, brittleness, elongation, flow characteristics, and viscosity of the mixture. Invented in 1953 by Karl Ziegler et al., (invention gave Ziegler a Nobel Prize in Chemistry), the high-density polyethylene (HDPE) is a compound of carbon and hydrogen polymer, belonging to the class of straight-chain polymers and unbranched (see Figure 4.7) and its chemical composition is highly resistant to shock and chemicals [46].

HDPE is strong and tough, four times less permeable than lower density relatives, and the chain can contain from 500,000 to 1,000,000 carbon units long, and it is a highly crystalline molecule (60–90%), thermoplastic polymer type, and viscoelastic nonlinear; the first allows you to merge with the application of sufficient heat and be shaped, molded, or extruded;

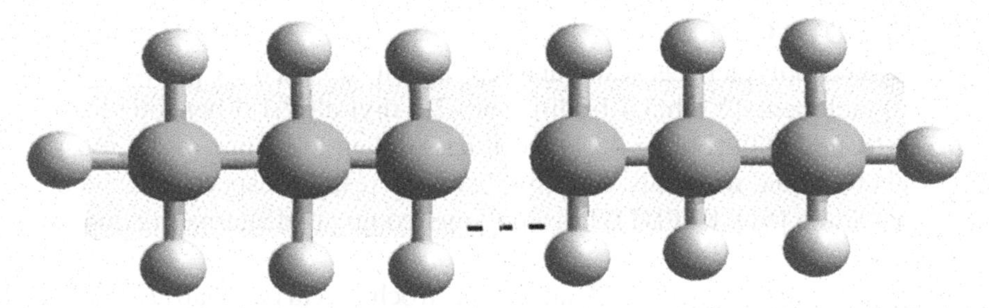

Figure 4.7 Structure of HDPE: carbon and hydrogen.

the second indicates that their properties depend on time. It resists most chemicals and solvents, and only a few substances can dissolve and this takes place at high temperatures; on the other hand, their chemical destruction can only occur under the action of strong oxidizing agents such as nitric acid vapor and sulfuric acid. In Table 4.1, other properties of HDPE are listed [46, 47].

Today, the HDPE is the polymer sold in the world and therefore the manufactured; more than half of its applications correspond to containers, caps, and closures, and the rest is distributed in toys, household accessories, packaging, sheets, pipes, among others. It has the advantage of being lightweight; flexible (even at low temperatures); elastic; resistant to corrosion, bacteria, chemicals, and water at 373 K; easy processing and transportation; and resistant to earthquakes, and it has long life and also be recyclable [48–50].

Table 4.1 Properties of HDPE

Properties	HDPE
Density	0.941–0.955 gm/cm^3
Flexural modulus (E_f)	$758 \leq E_f \leq 1103$ MPa
Tensile strength (F_t)	$21 \leq F_t \leq 24$ Mpa
Glass transition temperature	–243, –193 K
Melting point	408 K
Working temperature range (T_w)	$-173 < T_w < 393$ K
Volume resistivity	$>10^{15}$ Ω-cm
Surface resistivity	$>10^{15}$ Ω/square
Color	No color, Opaque
Smell	Odorless
Toxicity	Non toxic
pH	$1.4 < \text{pH} < 14$

4.1.3.2 PCPDTBT

The conjugated polymer has low bandgap, and it is called as poli[4,4-bis (2-etilhexil)ciclopenta[2,1-b;3,4-b′]ditiofeno-2,6-diyl-alt-2,1,3-benzotiadizol -4,7-diyl] (PCPDTBT) (Figure 4.8), and it has proven to be one of the most efficient photovoltaic materials with low bandgap response spectrum which extends to 900 nm [51]. PCPDTBT is a polymer conjugated alternating donor and acceptor.

It has a bandgap (E_g^{opt}) of 1.4 eV and an electrochemical bandgap (E_g^{echem}) of 1.7 eV with a HOMO level of –5.3 eV and a LUMO level of –3.6 eV. The range PCPDTBT response covers the entire range of visible light, making this feature the main advantage of this material, and it is also the first low bandgap polymer with high-efficiency photovoltaic response in the near-IR region [52–54].

It is noteworthy that the PCDTBT is amorphous and therefore it is not crystallized [55]. This polymer has a change of being in films solid to be in solution state; this is known as a phenomenon of solvatochromism (changes color depending on the polarity of the solvent); this behavior is observed in polymers π-conjugated strong actions inside and outside chain and indicates a 2D stack; that is, it indicates the extent of conjugation of more than two dimensions [56]. PCPDTBT luminescence can be almost completely cooled with an acceptor material, and a very high efficiency of luminescent sample cooling is very fast photoinduced charge transfer which is one of the main requirements for efficient solar cells. The other is the ability to transport individual mixing properties [57].

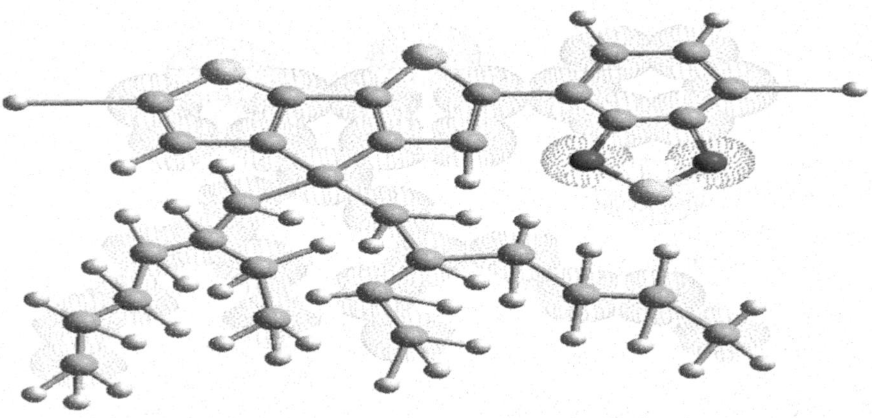

Figure 4.8 Chemical structure of PCDTBT: ● carbon, ● nitrogen, ○ hydrogen, and ○ sulfur.

In the unit cyclopenta [2,1-b; 3,4-b′] dithiophene (CPDT) of PCPDTBT, the two units of thiophene are confined to a plane, which extends the effective length of the polymer conjugate and effectively reduces the bandgap. Furthermore, the flat structure facilitates carrier transfer between two conjugated backbones, allowing mobility hollow PCPDTBT to achieve 1×10^{-3} cm^2/Vs, which is a rather high value for conjugated polymers. Furthermore, the two alkyl side chains in the polymer give the unit CPDT with excellent solubility, which is very important for uniform film in the process of spin-coating formation. Based on these properties, CPDT is a very useful unit in photovoltaic materials [53].

4.1.4 Graphite

Its name is derived from the Greek term, *graphein,* and refers to its use as a tool for writing. Natural graphite is an allotrope of carbon; in fact, graphite is the most stable form of carbon, as shown in Figure 4.9. It is found in nature in small hexagonal-shaped crystals compacted, scaly, earthy, and spherical aggregates [58].

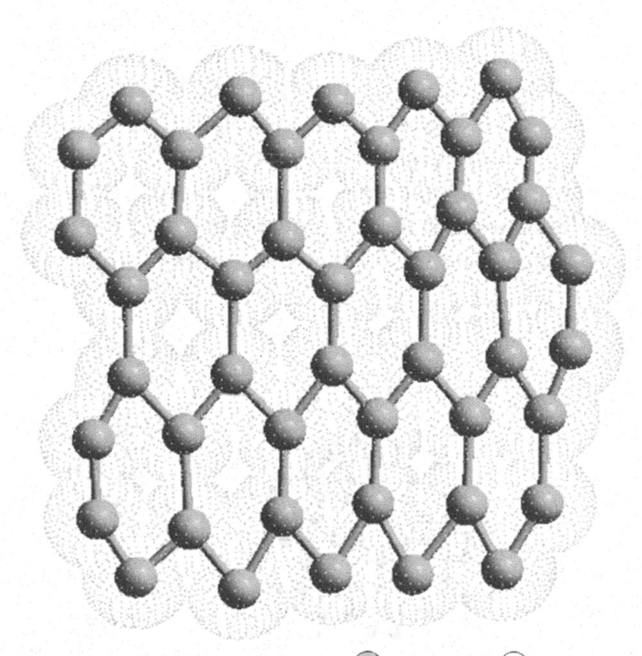

Figure 4.9 Structure of graphite: ● carbon and ○ hydrogen.

It is a soft mineral, gray to black, metalloid luster, has specific weight of 2.23, and chemically inert at normal temperature, odorless, non-toxic, heat resistant, and excellent conductor of heat and electricity. It is extremely refractory, being little affected by temperatures above 3,273 K; it has high resistance to acids and exposure to environmental agents; it is easily mixed with other materials, both liquid and solid [59]. It is compressible and pliable; it resists chemical attack, thermal shock, shrinkage, and oxidation; it has low coefficients of friction and thermal expansion; it is flexible and can easily be cut in a wide temperature range; and it is also an excellent lubricant.

Graphite or compounds of this can be obtained, and also other carbon allotropes materials, for example, graphite oxide, can be obtained, and by two bars of graphene graphite using a process called arc vaporization, the fullerene is obtained [60].

4.1.4.1 Fullerene

Fullerenes (named after the architect Buckminster Fuller) were accidentally discovered in 1985, although registration of their structures already had since the 1960s, since its properties have captivated the attention of scientists. Shaped molecule composed entirely enclosed cage of carbon atoms; therefore, the fullerene is a carbon allotrope, and graphite and diamond are shown in Figure 4.10 [61, 62]. The smallest possible fullerene is C20, which contains twelve pentagons and zero hexagon. Other possible fullerenes are C28, C32, C50, C60, C70, C80, C82, and C84.

Pentagons that are part of fullerenes are particularly stable, responsible for the closed structure, prevent the exchange edge, and that is where the molecular stress is concentrated. C60 has sixty carbon atoms which form

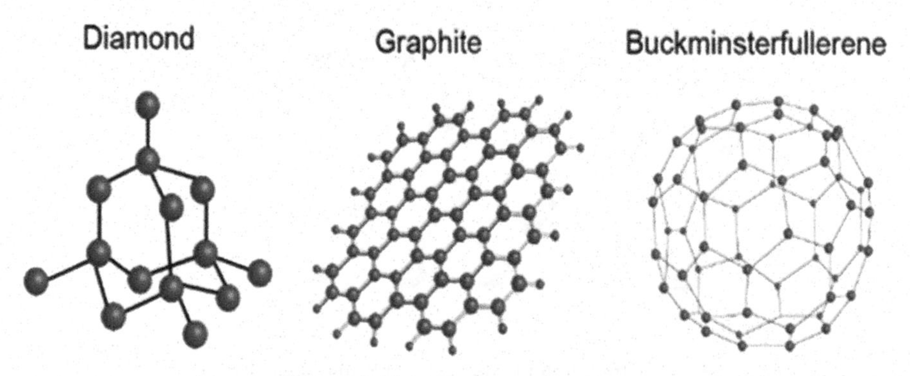

Figure 4.10 Allotropes of carbon.

twelve pentagons and twenty hexagons; each vertex is occupied by a carbon atom, where each carbon is bonded to three other carbons in an infinite two-dimensional array by a single bond and a double bond. The carbon atoms with this connectivity are usually called "Carbons sp^2," because in the three adjacent carbons, the orbitals used for sigma bonds are hybrid orbital 2s and two 2p orbitals (2px and 2py); the remaining 2p orbital is responsible π bond [63].

The most important property is its high symmetry C60. Chemically, the C60 molecule is very electronegative and readily forms compounds with electron donor atoms. Fullerenes have spheres of 7–15 Å diameter which is 6–10 times greater than the diameter of a typical atom. Fullerenes absorb strongly in the UV spectrum and moderately in the visible light. Theoretical calculations predict that this molecule has a relatively low LUMO energy. The bandgap of fullerenes is 1.68 eV [64].

At the atomic level, they are enormous, but they are small compared to many organic molecules. Due to their molecular character, fullerenes can be dissolved in the organic solvents and be chemically modified to lead to a high number of derivatives that generally retain exceptional physical and chemical properties of fullerene precursors. The larger size is smaller fullerene solubility and are completely insoluble in polar solvents or hydrogen bonds. The heat of formation is 10.16 Kcal/mol, which makes it less thermodynamically stable than the other allotropes of carbon. In Table 4.2, other properties of fullerenes are shown [65].

Fullerenes can be used as lubricants (chemically modified to contain another atom outside the sphere), superconductors, limiting liquids, liquid crystals, preparation of electronic devices, HIV inhibitor, powerful biological antioxidant, artificial photosynthetic systems, among others [63, 64].

Table 4.2 Properties of fullerenes C60

Physical Properties	Values
Density	1.65 gm/cm^3
Refractive index	600 nm
Tensile strength (F_t)	$21 \leq F_t \leq 24$ Mpa
Boiling point	Sublimes at 800K
Resistivity	1014 Ω/m
Pressure steam at 298K	5×10^{-6} torr
Pressure steam at 800K	8×10^{-4} torr
Organoleptic Properties	Values
Appearance	Solid black
Smell	Odorless

Fullerenes incorporating polymers would provide the polymer potentially most of the properties of fullerene. The film containing fullerenes formation is of great interest due to the possibility of transferring the properties of fullerenes by a simple coating process. Because of their physical and chemical properties of the possible applications of fullerenes, they are becoming more abundant and promising today [66].

4.1.5 HyperChem

Designed by Hypercube, software HyperChem is a sophisticated molecular modeling and simulation that allows complex chemical calculations. It is known for its quality, flexibility, and ease of use; it joins 3D visualization and animation with quantum chemical calculations, molecular mechanics, and dynamics. It incorporates programs such as *Ab-initio*, semiempirical molecular mechanics, and Monte Carlo simulation; in all components of the structure, thermodynamics and kinetics spectrum are included. With the various models, vibrational frequencies, transition states, and excited electronic states can be calculated [21, 67].

4.1.5.1 Calculation properties

The different conformations of a system and its coupling with the calculation methods allow the study of molecular properties dependent on conformation. The following properties can be calculated using HyperChem:

- **Geometry optimization:** Calculates the coordinates of a molecular structure representing a minimum of potential energy $\left(\frac{\partial V}{\partial \mathrm{r_i}} = 0\right)$. It identifies a minimum of potential energy to find a new stable structure before making other calculations that will provide a long set of electronic and structural properties and to prepare a molecule for molecular dynamics simulation [21, 68].
- **QSAR properties:** They are an effort to correlate the molecular structure and the properties derived from the molecular structure, with a particular type of chemical or biochemical activity. This type of activity is a function of interest of the person conducting the study; it is widely used in the pharmaceutical, environmental, and agricultural chemistry in search of particular properties. The calculations are empirical and therefore fast. The following properties can be calculated: partial atomic loads, surface area of van der Waals and solvent accessible, hydration energy, molecular volume of the solvent accessible surface delimited, Log P (measure of hydrophobicity), refractivity, polarizibility, and mass [21, 69–71].

- **FTIR:** Infrared Fourier transform spectrum is the preferred method of infrared spectroscopy; it is a technique that measures the vibration frequencies of the bonds in a molecule and provides a reflection spectrum bands of the functional groups of the inorganic and organic substances, which can make an identification of materials to determine the quality or consistency of a sample and identify unknown materials [72, 73].
- **MESP:** The electrostatic molecular potential is the electrical potential created by the distribution of electric charges such as electrons and the nucleus of a molecule; it correlates with dipole moment, electronegativity, and partial loads. It is useful because it helps in optimizing electrostatic interactions and indicates the reactive behavior toward charged species (places like charges repel and unlike charges attract). In electrostatic potential maps, the way they are distributed electrons in the molecule is shown usually using the colors of the rainbow (red is the area where there are more electrons and blue area with fewer electrons) [29, 74, 75].
- **UV spectroscopy:** It is based on the absorption process of ultraviolet–visible radiation by a molecule. The absorption of this radiation causes a promotion of an electron to a more excited state. Electrons which are excited by absorbing radiation at this frequency are the bonding electrons of the molecules, so absorption peaks can be correlated with different types of bonds present in the compound. It is commonly used in the quantitative determination of solutions of transition metal ions and highly conjugated organic compounds [76–79].

4.2 Methodology

4.2.1 Determination of Individual Structures

Individual molecules (Figure 4.11) will be determined by the software package HyperChem Pro 8.0, using the semiempirical method PM3. For calculating the properties of the individual components of molecules, first located in the toolbar (tool icons), and then, the structure is modified with the model build command to correct the bond distances, and angles of the draw command are used for the molecule for better appreciation of atoms and types of bonds.

4.2.2 Calculation of Energy

Once drawn the molecule, the PM3 method, which is to solve the Schrödinger equation to find the energy optimization, is used; geometry is optimized by

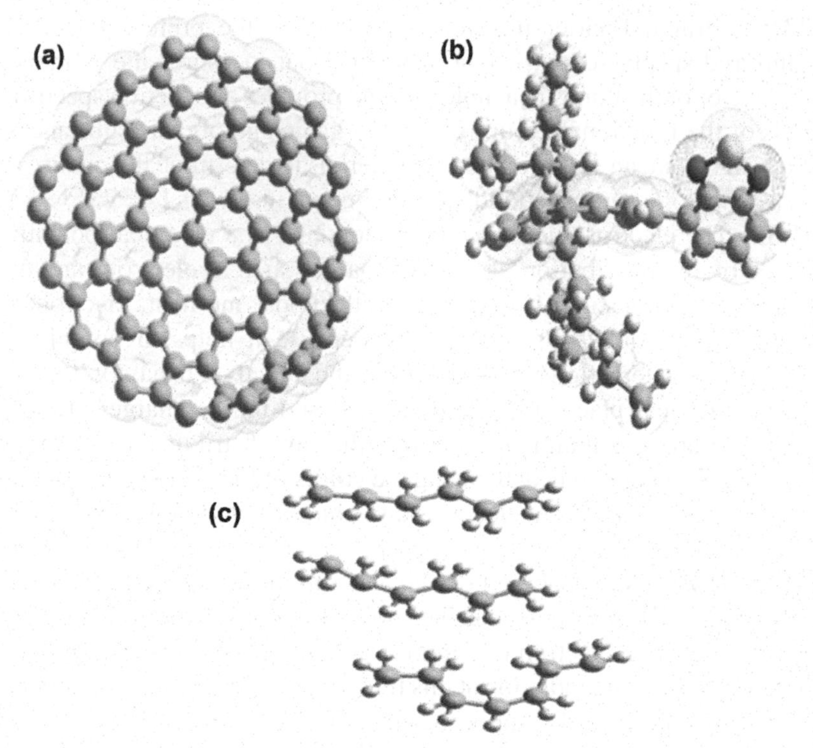

Figure 4.11 Structures of molecules: (a) fullerene, (b) PCPDTBT, and (c) LDPE, respectively.

selecting the Polak-Ribiere method for geometry optimization option compute tool. After the geometry is optimized, you can see the energy associated with each of the molecules in status line (bottom left of the screen).

4.2.3 Getting QSAR Properties

For the selection of properties QSAR (quantitative structure–activity relationships) such as partition coefficient, surface area, volume, and mass of each molecule, the compute/QSAR properties of the menu bar option are chosen.

4.2.4 Obtaining Electrostatic Potential Map

HyperChem shows the electrostatic potential map (MESP) to select the compute/plot molecular graphs of the menu bar option; later in the dialog window, molecular properties option is chosen.

4.2.5 Infrared Spectral Analysis (FTIR)

To determine the wavelengths of each molecule, the compute/vibration rotation analysis/vibrational spectrum is selected. Spectra were analyzed in various vibrations selecting a particular frequency and applying the animate vibrations bond type; corresponding vibrational mode displays its command, that way.

4.2.6 Obtaining Structural Parameters

The structural parameters are calculated by selecting build/constrain bond length and bond angle.

4.3 Results

4.3.1 Structural Analysis of PCPDTBT–Fullerene–Polyethylene

4.3.1.1 Energy optimization

The structural analysis of the composite was made with the corresponding minimum optimal ratio, that is, one molecule of each individual material. Energy optimization for the nanocomposite polymer matrix is given in terms of change in Gibbs free energy (ΔG); the result is –400,862 Kcal/mol, which is a negative amount, and this shows that the material is stable and has natural tendency to form without any external agent. In addition to the reaction releases energy (exothermic) thermodynamically, this is attributed to an increase in entropy or enthalpy decreased system [80, 81]. In Figure 4.12, you can see the energy corresponding to the geometry optimization and new links formed due to cross-linking in nitrogen to carbon PCPDTBT with fullerene and fullerene carbon–hydrogen for polyethylene.

4.3.1.2 Electrostatic potential map

The electrostatic potential map is shown in Figure 4.13; in it, the electronic distribution of the molecule is observed, with most of this red; this means that there is a greater probability of finding the electron in that area, the blue color represents the areas where there are fewer electrons and green neutral zones C = NS group is susceptible to attack an electrophilic attack, because of the electronegativity possessing these elements to be more electronegative nitrogen that carbon at the intersection formed between Fullerene and PCPDTBT, the latter who will cede electrons; no abrupt changes are seen in the colorations due to differences in electronegativity between the elements forming the

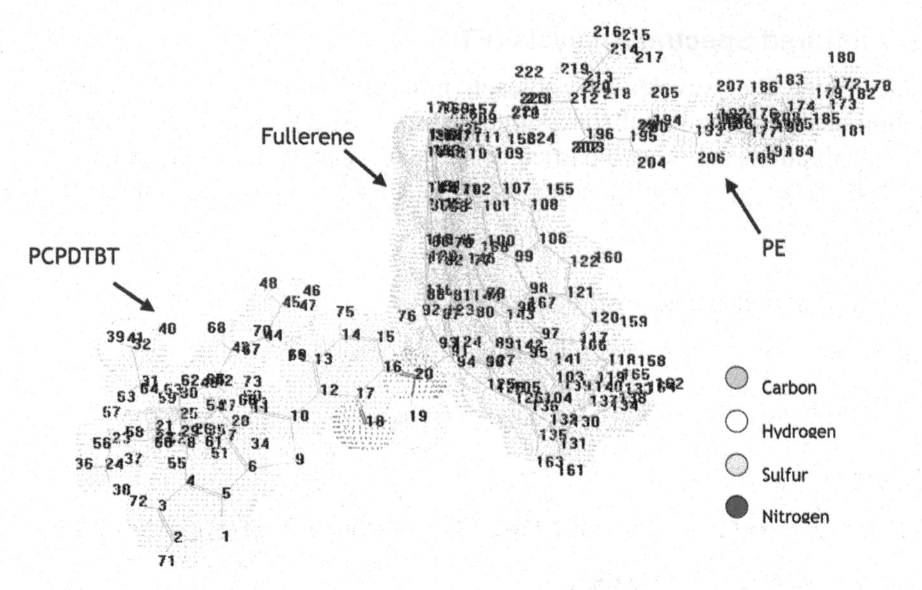

Figure 4.12 Molecule optimized with PM3.

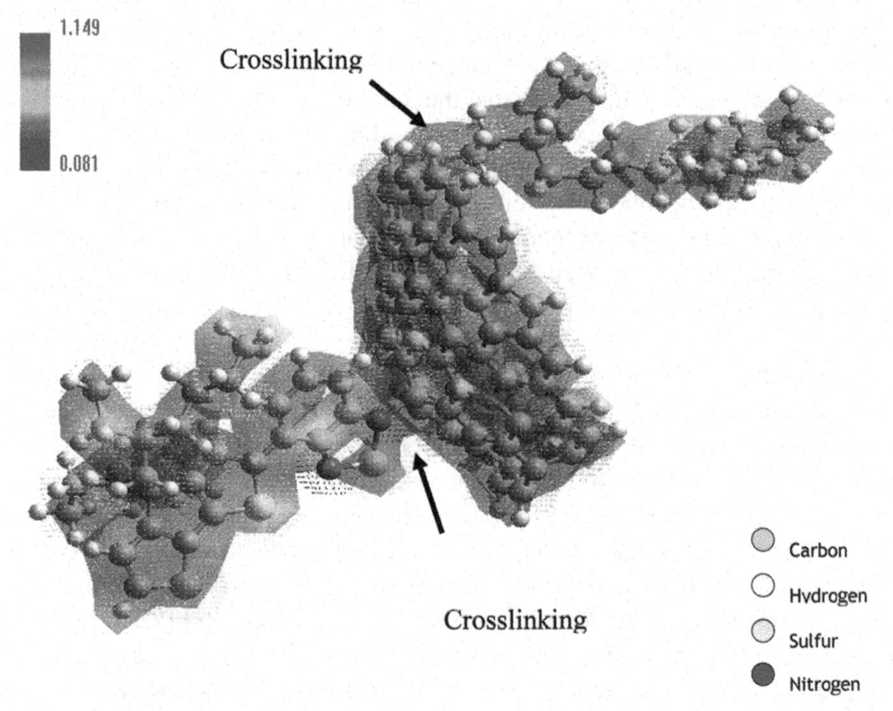

Figure 4.13 Map of electrostatic potential calculated with PM3.

material is not very large. In the case of cross-linking of fullerenes with polyethylene, the electronegativity difference indicates that hydrogen will use its attraction to form electron [82].

4.3.1.3 Bond length

Bond lengths presented in Tables 4.3–4.5 belong to each of the materials forming the composite material; although they are consistent with theoretical data, there are small variations that can be attributed to the formation of new bonds: carbon–nitrogen (C91-N20) and carbon–hydrogen (C153-H227), and in the case of carbon bound, compound length depends on the type of hybridization that has this element [83–85].

Table 4.3 Bond lengths obtained with PM3: PCDTBT

Bond	Length (Å)	Bond	Length (Å)	Bond	Length (Å)
S1–C2	1.7389	C16–C17	1.4603	C28–H33	1.0976
S1–C5	1.7191	C16=N20	1.3920	C28–H34	1.0982
C2–H71	1.0882	C17=N18	1.3450	C28–H35	1.0982
C2=C3	1.3861	N18–S19	1.7048	C29–H60	1.1095
C3–H72	1.0904	S19–N20	1.7849	C29–H61	1.1103
C3–C4	1.4109	N20–C91	1.5646	C29–C30	1.5397
C4=C5	1.4230	C21–H58	1.1137	C30–H62	1.1252
C4–C8	1.5170	C21–H59	1.1140	C30–C31	1.5316
C5–C6	1.4191	C21–C22	1.5345	C30–C42	1.5358
C6=C7	1.4284	C22–H55	1.1192	C31–H63	1.1100
C6–S9	1.7172	C22–C23	1.5359	C31–H64	1.1082
C7–C8	1.5153	C22–C25	1.5313	C31–C32	1.5131
C7–C11	1.3942	C23–H56	1.1121	C32–H39	1.0970
C8–C21	1.5341	C23–H57	1.1084	C32–H40	1.1032
C8–C29	1.5418	C23–C24	1.5115	C32–H41	1.0985
S9–C10	1.7663	C24–H36	1.0972	C42–H65	1.1086
C10=C11	1.4151	C24–H37	1.0782	C42–H66	1.1116
C10–C12	1.4316	C24–H38	1.1028	C42–C43	1.5219
C11–H73	1.1023	C25–H53	1.1092	C43–H67	1.1190
C12=C13	1.4043	C25–H54	1.1116	C43–H68	1.1135
C12–C17	1.4384	C25–C26	1.5211	C43–C44	1.5207
C13–H74	1.1081	C26–H51	1.1089	C44–H69	1.1086
C13–C14	1.3992	C26–H52	1.1136	C44–H70	1.1083
C14–H75	1.1028	C26–C27	1.5202	C44–C45	1.5138
C14=C15	1.3892	C27–H49	1.1082	C45–H46	1.0976
C15–H76	1.0969	C27–H50	1.1082	C45–H47	1.1059
C15–C16	1.4147	C27–C28	1.5123	C45–H48	1.0979

Table 4.4 Lengths obtained with PM3: fullerene

Bond	Length (Å)	Bond	Length (Å)	Bond	Length (Å)
C77–C78	1.4198	C100–C101	1.4177	C127–C128	1.4234
C77–C79	1.4301	C101–C102	1.4186	C127–C142	1.4387
C77–C100	1.4270	C101–C107	1.4353	C128–C136	1.4235
C78–C82	1.4264	C102–C110	1.4251	C129–C145	1.4424
C78–C83	1.4325	C103–C104	1.4089	C130–C131	1.4437
C79–C80	1.4191	C103–C119	1.4419	C130–C134	1.4441
C79–C96	1.4288	C104–C105	1.4315	C131–H161	1.0848
C80–C81	1.4303	C104–C130	1.4280	C131–C132	1.4791
C80–C89	1.4359	C105–C126	1.4282	C132–C135	1.4799
C81–C82	1.4390	C106–C108	1.4139	C132–C137	1.5702
C81–C87	1.4226	C106–C122	1.4234	C133–H162	1.0935
C82–C86	1.4230	C107–C108	1.4259	C133–C134	1.4262
C83–C84	1.4220	C107–C109	1.4261	C134–C138	1.0951
C83–C102	1.4299	C108–H155	1.0945	C135–H163	1.0850
C84–C85	1.4235	C109–H156	1.0959	C135–C136	1.4458
C84–C148	1.4322	C109–C111	1.3835	C136–C139	1.4408
C85–C86	1.4269	C110–C111	1.4276	C137–C138	1.4264
C85–C112	1.4219	C110–C147	1.4090	C137–C140	1.4211
C86–C114	1.4195	C111–H157	1.0960	C138–H164	1.0951
C87–C88	1.4249	C112–C113	1.4248	C139–C140	1.4357
C87–C91	1.5275	C112–C150	1.4342	C139–C141	1.4243
C88–C92	1.4092	C113–C115	1.4153	C140–H165	1.0953
C88–C114	1.4351	C113–C151	1.4427	C141–H166	1.0944
C89–C90	1.4223	C114–C115	1.4300	C141–C142	1.4401
C89–C95	1.4374	C115–C129	1.4272	C142–C143	1.4200
C90–C91	1.5135	C116–C123	1.4521	C143–H167	1.0954
C90–C105	1.4295	C116–C129	1.3811	C143–C144	1.4360
C91–C94	1.5224	C117–C118	1.4173	C144–C146	1.4650
C92–C93	1.4365	C117–C120	1.4486	C145–C146	1.4011
C92–C116	1.4325	C118–H158	1.0959	C145–C152	1.4311
C93–C94	1.4007	C118–C119	1.4167	C146–H168	1.0947
C93–C124	1.4463	C119–C133	1.4273	C147–H169	1.0965
C94–C125	1.4295	C120–H159	1.0951	C147–C148	1.4160
C95–C97	1.4185	C120–C121	1.3745	C148–C149	1.4193
C95–C103	1.4104	C121–C122	1.5053	C149–H170	1.0955
C96–C97	1.4123	C122–H160	1.0859	C149–C150	1.4035
C96–C98	1.3920	C123–C124	1.4275	C150–C154	1.3973
C97–C117	1.4342	C123–C144	1.3800	C151–C152	1.4104
C98–C99	1.4095	C124–C127	1.3885	C152–H171	1.0968
C98–C121	1.4291	C125–C126	1.4212	C153=C154	1.3600
C99–C100	1.3952	C125–C128	1.4399	C153–H227	1.0995
C99–C106	1.4399	C126–C132	1.5578		

Table 4.5 Lengths obtained with PM3: polyethylene (PEHD)

Bond	Length (Å)	Bond	Length (Å)	Bond	Length (Å)
Chain: 1		Chain: 2		Chain: 3	
C172–H178	1.0968	C209–H225	1.0878	C191–H197	1.1087
C172–H179	1.1032	C209–H226	1.0825	C191–H208	1.1155
C172–H180	1.0977	C209–C227	2.7857	C191–C192	1.5196
C172–C173	1.5129	C209–C210	1.4606	C192–H198	1.1129
C173–H181	1.1082	C210–H223	1.1102	C192–H207	1.1088
C173–H182	1.1087	C210–H224	1.1173	C192–C193	1.5196
C173–C174	1.5344	C210–C211	1.5226	C193–H199	1.1089
C174–H183	1.1186	C211–H221	1.1089	C193–H206	1.1090
C174–C175	1.5334	C211–H222	1.1081	C193–C194	1.5195
C174–C191	1.5327	C211–C212	1.5299	C194–H200	1.1080
C175–H184	1.1091	C212–H220	1.1205	C194–H205	1.1168
C175–H185	1.1101	C212–C213	1.5348	C194–C195	1.5205
C175–C176	1.5207	C213–H218	1.114	C195–H201	1.1142
C176–H186	1.1082	C213–H219	1.1071	C195–H204	1.1076
C176–H187	1.1127	C213–C214	1.5135	C195–C196	1.5195
C176–C177	1.5118	C214–H215	1.0968	C196–H202	1.1091
C177–H188	1.0975	C214–H216	1.0977	C196–H203	1.1141
C177–H189	1.0981	C214–H217	1.1053	C196–C212	1.5318
C177–190	1.0981				

4.3.1.4 Spectrum Fourier Transform Infrared (FTIR)

The signals observed in the infrared spectrum are shown in Table 4.6, and corroborate the composition of the composite material, these signals are within or very close to the frequencies reported in bibliographic data; it shows the stability of this material. Theoretical vibrations for CH group can be found in the region of 3000–1500 cm^{-1}: for DC 1600–500 cm^{-1}, C = C 1700−1500 cm^{-1}, C = N (1690−1480 cm^{-1}), CN (1250−1020 cm^{-1}), CS (710−570 cm^{-1}), and SN ∼700 cm^{-1}. You can also observe the vibrations that belong to the new bonds between materials which are the product of crossovers. The fact that these signals appear in the infrared spectrum confirmed that there exists cross-linking of the individual materials to form the composite [86−88].

Table 4.6 FTIR results for PCDTBT/F/PE

Type of Vibration	AMBER	PM3
C–H stretching	5661, 5580, 5578, 5651	3152–3023, 2915–2540, 2362–2128, 1967, 1707–1416, 1227, 1117
C–H asymmetric stretching	5653, 5568, 5553, 5552	2896–2876, 2545–2119, 2099–2030, 1938–1726, 1521–1332
C–H symmetric stretching	5501, 5437, 5424, 5420	3068–2965, 2703–2606, 2568–2509, 2465–2268, 1487
C–H scissoring	2794–2701, 2686–2683	————
C–H torsion	2650, 2399, 2213–2202, 2181–2105, 2039	————
C–H swinging	2610, 2519, 2514, 1642–1622, 1548, 1493, 1344	1051
C–H swinging	2610–2514, 1642–1622, 1548–1327	1051
C–C stretching	3212, 3207, 3003, 2973–2901, 2889–2861, 1805	1704, 1691–1604, 1596–1507, 1498–1405, 1398–1305, 1297–1200, 1192–1100, 1086–1023, 998–903, 896–842, 742–702, 690–630, 578–543, 473–404, 364–309, 191–146
C=C stretching	3675–3559, 3498–3404, 3384, 2689–2476, 1604	1683, 1558–1507, 1487–1405, 1398–1305, 1297–1200, 1192–1100, 1068–1031, 962–903, 896–842, 630
C=C flexion (Al, Ti, To, Ba)	1672–1596, 1478–1468, 1206–1099, 956–952,	690
S–C	1727–1711, 1185–1133, 915, 844	962–903, 873–822, 789–702, 647
N=C	2794, 2672–2303, 1664–1478, 742	1643–1633, 1596–1507, 1487–1405, 1390, 1264–1243, 1166–1100, 1051–1031, 873–822, 789
N–S	1722–1185, 803, 775–750, 685–599	1524, 990–903, 873–857, 789–702, 690–630, 404
N–C	3075, 3015, 2690–2506	2990, 1050, 822
C–H	4497–4474, 2481–2334, 1251	2979–2807, 2789

4.4 Conclusions

The structural analysis of composite PCPDTBT–fullerene–polyethylene was performed using a hybrid of quantum mechanics (AMBER–PM3) and molecular mechanics, and thus, the properties presented in the results are the Gibbs free energy, electrostatic potential map obtained, bond lengths, and infrared spectrum Fourier transform (FTIR) of each we can conclude the following:

- **Gibbs free energy (ΔG):** Individual materials react spontaneously, the reaction is exothermic, and the formed composite is stable. That is why PM3 results in a negative energy optimization for our system.
- **Electrostatic potential map:** The molecule may be susceptible to electrophilic attack, and the change in color between nitrogen and carbon–PCPDTBT–fullerene indicates an attraction and therefore intersecting between the materials; same situation occurs between a fullerene–carbon with hydrogen polyethylene.
- **FTIR:** The main signals verifying the composition of our material were obtained, the abundance of the number of signals corresponding to the large number of bonds present in the composite, and the appearance of the signs of the new formed bonds confirms the existence of cross-linking the materials.

Acknowledgments

Thank God for giving me everything without merit anything. Dra. Norma Aurea Rangel Vázquez thanks for your incredible patience and do our job the best job you could hope for and for taking the time to share their knowledge with me. Thanks Mom for being the engine that drives all, the mood for the days where it seems that everything goes wrong and happier person when things go well. Thanks Daddy because you are there when I need support and who discuss my ideas or when I need advice. Julito, thanks for being very smart and at the same time loving me so much: you are my favorite brother.

References

[1] Clementi, E. (2012). Evolution of Computational Chemistry, the "Launch Pad" to Scientific Computational Models: the Early Days from a Personal Account, the Present Status from the TACC-2012 Congress, and Eventual Future Applications from the Global Simulation Approach. In J.-M. A. Enrico Clementi, *Theory and Applications in Computational*

Chemistry: The First Decade of the Second Millennium (pp. 5–54). (Como: American Institute of Physics).

[2] Taylor, G. (2012). The Next Decade of Computing. In J.-M. A. Enrico Clementi (ed.), *Theory and Applications in Computational Chemistry: The First Decade of the Second Millennium*, pp. 55–57. (Hillsboro: American Institute of Physics).

[3] Piela, L. (2012). From Quantum Theory to Computational Chemistry. A Brief Account of Developments. In J. Leszczynski (ed.), *Handbook of Computational Chemistry*, pp. 2–11. (Warsaw: Springer Science+Business Media B.V.).

[4] Jensen, F. (2007). *Introduction to Computational Chemistry.* (Chichester: John Wiley & Sons Ltd.).

[5] Lewars, E. G. (2011). *Computational Chemistry. Introduction to the Theory and Applications.* (Peterborough: Springer Science+Business Media B.V.).

[6] Young, D. C. (2001). *Computational Chemistry: A practical Guide for Applying Techniques to Real-World Problems.* (New York: John Wiley & Sons, Inc.)

[7] *Introducción a la Mecánica cuántica.* (19 de February de 2016). Obtenido de Tripod: http://lqi.tripod.com/Modelado/MECMOL.HTM

[8] Perez, M. S. (1994). *Mecánica Molecular.* Madrid: Departamento de Química Orgánica, Universdad de Alcalá.

[9] Sherrill, C. D. (2001). *Introduction to Molecular Mechanics.* Georgia: School of Chemistry and Biochemistry Georgia Institute of Technology.

[10] Sanger, A. (2007). *"LAS FUERZAS Y SU MEDICIÓN": LEY DE HOOKE.* Villa Eloisa: Instituto Balserio.

[11] Shattuck, T. W. (2008). *Molecular Mechanics Tutorial.* Maine: Department of Chemistry Colby College.

[12] Hehre, W. J. (2003). *A Guide to Molecular Mechanics and Quantum Chemical Calculations.* Irvine: Wavefunction, Inc.

[13] Mayorga, O. L. (31 de August de 2010). *Interacciones por enlaces o puentes de hidrógeno.* Obtenido de Estructura de macromoléculas: http://www.ugr.es/~olopez/estruct_macromol/fuerzas/EH.PDF

[14] Deriabina, J. L. (2006). Desarrollo de un campo de fuerzas de mecánica molecular para la interacción de Na+ con agua. *Revista Mexicacna de Física*, 75.

[15] Kenno Vanommeslaeghe, O. G. (2015). Molecular mechanics. *Current Pharmaceutical Design.*

[16] Rodrigo Galindo-Murillo, T. E. (2016). Using information about DNA structure and dynamics from experiment and simulation to give insight into genome-wide association studies. En A. Rodríguez-Oquendo, *Translational Cardiometabolic Genomic Medicine,* p. 85. (USA: Elsevier).
[17] D. A. Case, R. B. (2016). *AMBER 16 Reference Manual.* San Francisco: Universidad de California.
[18] Wendy, D., and Cornell, P. C. (1995). A second generation force field for the simulation of proteins nucleic acids, and organic molecules. *J. Am. Chem. Soc.*, 5179–5184.
[19] Scott, J., and Weiner, P. A. (1986). An all atom force field for simulations of proteins and nucleic acids. *J. Comput. Chem.* 230, 231.
[20] Scott, J., and Weiner, P. A. (1984). A new force field for molecular mechanical simulation of nucleic acids and proteins. *J. Am. Chem. Soc.*, 765–769.
[21] Hypercube Inc. (2002). *HyperChem Manual Release7.* USA: Hypercube Inc.
[22] Vieyra, J. C. (30 de July de 1997). *Introducción a la Mecánica Cuántica.* Obtenido de Instituto de Ciencias Nucleares de la UNAM: http://www.nucleares.unam.mx/~vieyra/node23.html#SECTION00014 0000000000000000
[23] Enciclopedia Contributors. (11 de July de 2010). *Mecánica Cuántica.* Obtenido de Enciclopedia, De la Enciclopedia Libre Universal en Español.: http://enciclopedia.us.es/index.php–title=Mec%C3%A1nica_cu%C3%A1ntica&oldid=522217
[24] Gratton, J. (2003). *Introducción a la Mecánica Cuántica.* Buenos Aires: Instituto de Física de Plasma.
[25] Stewart, J. J. (2016). *Semiempirical Theory—Introduction.* Obtenido de Theory used in MOPAC2016: http://openmopac.net/manual/semiempiric al_theory.html
[26] Kahn, K. (2007). *Semiempirical Quantum Chemistry.* Obtenido de Department of Chemistry and Biochemistry: http://people.chem.ucsb.edu /kahn/kalju/chem226/public/semiemp_intro.html
[27] Thiel, W. (2005). Semiemprical quantum-chemical methods in computational chemistry. In: En C. D. et al. (eds), *Theory and Applications of Computational Chemistry: The First Forty Years*, pp. 563–566. (Indianapolis: Elsevier B.V.).
[28] Thiel, W. (2013). Semiempirical quantum–chemical methods. *Comput. Molecul. Sci.* 145–157.

[29] Rangel-Vázquez, N.-A. (2015). Importance of molecular simulation for studying structural properties. *Mater. Sci. Eng. Adv. Res. J.* 1,3,4.

[30] Steward, J. J. (1989). Optimization of parameters for semiempirical methods I. methods. *J. Comput. Chem.* 209–220.

[31] Daniel Gay, S. V. (2003). *Composite materials: design and applications.* (Boca Raton: CRC Press LLC).

[32] Halpin, J. C. (1992). *Primer on composite materials analysis.* (Lancaster: Technomic Publishing Company, Inc).

[33] Barbero, E. J. (2011). *Introduction to composite materials design.* (Boca Raton: CRC Press, LLC).

[34] F. L. Matthews, R. D. (1999). *Composite materials: engineering and science.* (Boca Ratón: CRC Press LLC).

[35] Ru-Min Wang, S.-R. Z. (2011). *Polymer matrix composites and technology.* (Sawston: Woodhead Publishing).

[36] Wolff, E. G. (2004). *Introduction to the dimensional stability of composite material.* (Lancaster: DEStech Publications, Inc.)

[37] Jones, R. M. (1999). *Mechanics of composite materials.* (Philadelphia: Taylor & Francis, Inc.).

[38] Pedro Henrique Cury Camargo, K. G. (2009). Nanocomposites: synthesis, structure, properties and new application opportunities. *Mater. Res.* 1, 25, 26.

[39] Gerardo Martínez, H. S. (2010). *Nanocompuestos poliméricos a partir de grafeno.* Madrid: Instituto de Ciencia y Tecnología de Polímeros.

[40] D.R. Paul, L. R. (2008). Polymer nanotechnology: Nanocomposites. *Polymer*, 12–14.

[41] Farzana Hussain, M. H. (2006). Review article: polymer-matrix nanocomposites, processing, manufacturing, and application: an overview. *J. Comp. Mater.* 1512–1517.

[42] NetComposites Ltd. (2016). *Polymer Composites.* Obtenido de NetComposites: http://www.netcomposites.com/guide-tools/guide/introduction/polymer-composites/

[43] Drzal, L. T. (1989). The effect of polymeric matrix mechanical properties on the fiber-matrix interfacial shear strength. *Mater. Sci. Eng.* 289–293.

[44] U. S. Congress, Office of Technology Assesment. (1988). *Advanced materials by design.* (Washington, DC: U.S. Government Printing Office).

[45] Åström, B. T. (1997). *Manufacturing of Polymer-Matrix Composites.* (Cheltenham: Nelson Thornes Ltd.).

[46] Gabriel, L. H. (2003). History and physical chemistry of HDPE. In Lester O. B., Gabriel, H. (eds.) *Corrugated polyethylene pipe design manual & installation guide,* pp. 2–18. (Irving: Plastic Pipe Institute. Obtenido de Pars Ethylene Kish Co.).

[47] Wade, L. (2011). Polímeros sintéticos. In Wade, L. (ed) *Química Orgánica,* pp. 1236–1237. (Mexico: Pearson Education).

[48] The Editors of Encyclopædia Britannica. (2016). *Polyethylene (PE).* Obtenido de Encyclopædia Britannica: http://global.britannica.com/science/polyethylene

[49] Cornelia Vasile, M. P. (2005). *Practical guide to polyehylene.* (Shrewbury: Rapra Technology Ltd.).

[50] Peacock, A. J. (2000). *Handbook of polyethylene: structres, properties and applications.* (New York. Basel: Marcel Dekker, Inc.)

[51] Jianhui Hou, T. L.-Y. (2009). Poly[4,4-bis(2-ethylhexyl)cyclopenta[2,1-b;3,4-b#]dithiophene-2,6-dyil-alt-2,1,3-benzoselenadiazole-4,7-diyl], a New Low Band Gap Polymer in Polymer Solar Cells. *J. Phys. Chem C.*, 1601, 1602.

[52] Qiao, Q. (2015). *Organic solar cells: Materials, devices, interfaces and modeling.* (New York: Taylor & Francis Group).

[53] Steve Albrecht, S. J. (2012). Flourinated copolymer PCPDTBT with enhanced open-circuit voltage and reduced recombination for highly efficient polymer solar cells. *J. Am. Chem. Soc.* 14933.

[54] Christina Scharsich, F. S. (2015). Revealing structure formation in PCPDTBT by optical spectroscopy. *J. Polym. Sci.* 1416, 1417.

[55] Cauble, G. D. (2011). *Morphology changes in PCPDTBT: PCMB and P3HT: PCPDTBT: PCBM and its effect on polymer solar cell performance.*

[56] Waters, R. H. (2015). *Characterisation and lifetime studies of CPDT- and BT-based photovoltaic cells.* Bangor: Bangor University.

[57] David Mühlbacher, M. S. (2006). High photovoltaic performance of a low-bandgap polymer. *Adv. Mater.* 2884–2886.

[58] Asturnatura. (30 de May de 2016). *Grafito.* Obtenido de asturnatura: http://www.asturnatura.com/mineral/grafito/96.html

[59] Dirección General de Promoción Minera. (October de 2005). *Ficha Técnica.* Obtenido de Perfil de Mercado de la Grafito: http://201.131.19.30/estudios/mineria/sistema%20mineria/grafito/CARACTERISTICAS.htm

[60] Charles M., and Lieber, C.-C. C. (1994). Preparation of fullerenes and fullerene-based materials. In: Henry Ehrenreich, F. S. (ed.), *Solid state physics,* pp. 112–113. (Cambridge: Academic Press Inc.).
[61] Franco, M. Á. (2009). De la Química Interestelar al Nanocoche: Fullerenos y Nanotubos. *Real Academia de Ciencias Exactas, Físicas y Naturales*, 4, 6.
[62] Torres, T. (23 de Octubre de 2015). *Fullerenos y Nanotubos.* Obtenido de http://depa.fquim.unam.mx/amyd/archivero/Tomastorres7_27202.pdf
[63] Laowachirasuwan, K. (2008). Fullerenes and fullerene derivatives in properties and potential applications. *University of the Thai Chamber of Commerce J.* 166–168.
[64] Oxana Vasilievna Kharisssova, U. O. (2002). La Estructura del Fullereno C60 y sus Aplicaciones. *Ciencia UANL,* 2–4.
[65] Yadav, R. K. (2008). Structure, properties and applications of fullerenes. *Intl. J. Nanotechnol. Appl.* 20.
[66] León, N. M. (1999). Fullerenos: moléculas de carbono con propiedades excepcionales. *Anales de la Real Sociedad Española de Química Segunda Época*, 1–3.
[67] MakoLab. (2007). *HyperChem Professional 8.0.* Obtenido de Hypercube, Inc.: http://www.hyper.com/_tabid=360
[68] The Shodor Education Foundation Inc. (2000). *Background Reading for Geometry Optimizations.* Obtenido de ChemViz: https://www.shodor.org /chemviz/optimization/students/background.html
[69] Kubinyi, H. (1997). QSAR and 3D QSAR in drug design Part 1: methodology. *Drug Discovery Today*, 457.
[70] Jorge Lozano-Aponte, T. S. (2012). ¿Qué sabe Ud. acerca de ... QSAR? *Revista Mexicana de Ciencias Farmacéuticas*, 82–83.
[71] Fabiana Maguna, N. O. (2011). Relaciones cuantitativas estructura-actividad de compuestos antimicrobianos. *Reunión de Difusión de las Labores Docentes, Científicas y Tecnológicas, y de Extensión*, 1.
[72] Sil, J. L. (2016). *Espectrometría Infrarroja con Transformada de Fourier (FTIR).* Obtenido de Análisis No Destructivo para el Estudio in situ del Arte, la Arqueología y la Historia: http://www.fisica.unam.mx/andreah/tecnicas_equipos/ftir.html
[73] Nicolet, T. (2001). *Introduction to Fourier Transform Infrared Specctroscopy.* Obtenido de Molecular Materials Research Center: http://mmrc.caltech.edu/FTIR/FTIRintro.pdf

[74] Ophardt, C. E. (2003). *Molecular Electrostatic Potential.* Obtenido de Virtual Chembook: http://chemistry.elmhurst.edu/vchembook/211elecpo tential.html

[75] Manzo, R. A. (8 de March de 2010). *Mapas de Potencial Electrostático.* Obtenido de SlideShare: http://es.slideshare.net/xiuhts/mapas-de-poten cial-electrosttico

[76] *UV-vis Absorption Spectroscopy.* (17 de May de 2016). Obtenido de Sheffield Hallam University: http://teaching.shu.ac.uk/hwb/chemistry/tu torials/molspec/uvvisab1.html

[77] Bacher, A. D. (17 de May de 2016). *Theory of Ultraviolet-Visible (UV-Vis) Spectroscopy.* Obtenido de http://www.chem.ucl a.edu/~bacher/UV vis/uv_vis_tetracyclone.html

[78] Clark, J. (May de 2016). *UV-VISIBLE ABSORPTION SPECTRA.* Obtenido de Chemguide: http://www.chemguide.co.uk/analysis/uvvisibl e/theory.html

[79] Reusch, W. (05 de May de 2013). *Visible and Ultraviolet Spectroscopy.* Obtenido de Michigan State University Department of Chemistry: http://www2.chemistry.msu.edu/faculty/reusch/virttxtjml/spectrpy/uv-vis/spectrum.htm

[80] Atkins, P. (2007). *Four laws that drive the universe.* (Oxford: Oxford University Press).

[81] American Chemical Society. (2007). *Chemistry. A project of the American Chemical.* (New York: W. H. Freeman and Company).

[82] Norma-Aurea Rangel-Vázquez, F. R. (2014). *Computational chemistry applied in the Analises of Chitosan/Polyvinylpyrrolidone/Mimosa Tenuiflora.* (San Francisco: Science Publishing Group).

[83] Cartmell, B. S. (2003). *Valencia y estructura molecular.* (Barcelona: Revertré).

[84] David Gutsche, D. J. (1975). *Fundamentals of Organic Chemistry.* Englewood Cliffs: Prentice-Hall, Inc.

[85] Geissman, T. A. (1974). *Princios de Química Orgánica.* Barcelona: Revertré.

[86] Thomas J., and Bruno, P. D. (2011). *Handbook of basic tables for chemical analysis.* (Boca Raton: CRC Press).

[87] Carissa Hampton, D. D. (2010). *Vibrational Spectroscopy Tutorial: Sulfur and Phosphorus.* Obtenido de Show me Chemistry. The Glaser Group: https://faculty.missouri.edu/~glaserr/8160f10/A03_Silver.pdf

[88] Robert M., and Silverstein, F. X. (2005). *Spectrometric Identification of Organic Compounds.* (USA: John Wiley & Sons, Inc.)

5

Determination of Thermodynamic Properties of Ionic Liquids Through Molecular Simulation

Claudia-Lizeth Salas-Aguilar

Chemical Engineering Department,
PhD in Engineering Sciences,
Technological Institute of Aguascalientes, Mexico

Abstract

In the world of chemical, experimental thermodynamic data are essential in the design, optimization, simulation, and process control. Hence, the knowledge of the thermodynamic properties is essential in any activity related to the design and optimization. However, the thermodynamic properties of substances are often easy due to a variety of causes such as working with hazardous substances, to extreme conditions or having the high cost collection. Therefore, the search for alternatives that help us determine these thermodynamic data is needed. The molecular simulation is a good choice for a safe and less expensive manner the behavior of matter in different means. In this chapter, the methodology for the determination of the thermodynamic properties of ionic liquid 1-n-butyl-3-methyl imidazolium hexafluorophosphate (cation $[bmim]^+$ and anion $[PF6]^-$) by molecular simulation is thc Monte Carlo technique. Similarly, the models for estimating the Henry constant CO_2-ionic liquid, conditions, and modeling system are presented. In addition, the effect of the molecular description of the system ("all atoms" (AA) or "atoms bonded" (UA)) on the thermodynamic properties obtained is analyzed.

Keywords: Molecular simulation, thermodynamic properties, ionic liquids.

5.1 Introduction

In the final years of ionic liquids (ILs), they have attracted considerable interest in both the world of scientific research and in various technological and industrial sectors. The various interesting properties possessed these fascinating compounds such as low volatility, low melting point, high thermal stability, and favorable solvation behavior have made them a key to many fields of science element, among them green chemistry, electrochemistry, and nanomaterials.

The possibility of combining different organic cations, and generally large, with various inorganic anions usually arises the term *solvents design*, since binding of each of these ions that determine the physicochemical properties will possess the IL obtained. Among the most used for creating design of these solvents, cations are the families of the imidazolium, pyridinium, and the tetraalkyl phosphonium whose partners are usually halide anions, sulfates, sulfonates, amides, borates, and phosphates. These *solvents designs* have unique properties that make them attractive for various applications such as solvents in organic and inorganic compounds, nonvolatile compounds in chemical reactions, catalysis, hydrogenation, carbonylations, and oxidations aerobic.

However, one of the barriers that arise in the adoption of these fabulous ILs is limited data on their physicochemical properties. The lack of fundamental understanding of how these properties depend on the chemical constitution of the IL, along with great difficulty and cost of obtaining, makes prediction methods and computer simulation the best tool for estimating their properties.

Computer simulation and molecular simulation (MS) were presented as a tool for the prediction of properties of matter, basically for the study of liquid giving credibility and rapidly extending their use to subjects as diverse as quantum mechanics, the physics of fluids, plasma physics, condensed matter, nuclear physics, and materials science.

Thanks to the rapid development of computer technology, computer simulation has become an essential tool calculation since by good computational model can reproduce laboratory experiments also because they can be used freely vary the parameters, can prove or disprove existing theoretical models in ranges of parameters to achieve experimentally impossible for now, resolving conflicts between theoretical explanation and observation.

A key role is played also today viewing the results, because not only can obtain numerical data, also a graphic image of the process. In the present day, MS techniques are not only a useful tool for understanding and predicting the behavior of substances from a description at the molecular level, also can predict structural, thermodynamic, and transport properties.

5.1.1 Overview of Simulation

The MS methods can provide information to validate and analyze theoretical models. The molecular motions to simulate a system may generate information that cannot be measured directly in a laboratory, thereby enabling detailed study of various phenomena of substances. There are two main techniques for estimating MS physical and chemical properties of substances: They are the techniques of molecular dynamics (MD) and Monte Carlo (MC).

The MD technique is used to simulate fluid structure and molecular motion by solving the equations of motion of Newton, obtaining dynamic properties of the system, while MC uses a stochastic method for generating random molecular motions through displacement, where it is possible to simulate static properties of the system under study. Basically, MD and MC techniques have been used to simulate structural, thermodynamic, mechanical, and kinetics properties of gases, liquids, and solids.

To conduct a MS is necessary to design an initial system configuration. In the MC technique, positions and molecular orientation must be specified. In this technique, the simulations are performed in two stages.

The first stage is known as equilibration stage in which a number of configurations such that the system evolves toward equilibrium, that is, the properties fluctuate around a constant value. At the end of this period, begins the second stage known as production stage where the information necessary for calculating properties is studied. Likewise, parameters such as force field, the choice of collective simulation, and system conditions should take into account.

5.1.2 Implementation of the Simulation Method

A simulation for both MD and MC involves the development of a computer program as shown in Figure 5.1.

- Initialization: Once the choice of the group simulation and interaction potential is made, a molecular description of the system is done by specifying the initial conditions, as positions of the particles, temperature, volume, density, etc.
- Generation of configuration/simulations: MD is obtained by integrating the classical equations of motion of the system. In MC, they are obtained by a method of stochastic nature.

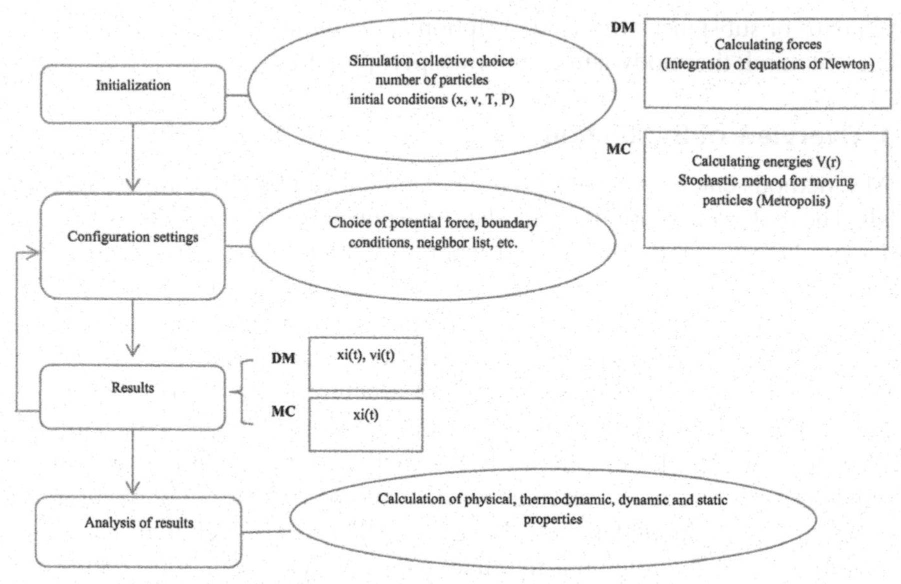

Figure 5.1 Central elements of a typical computer simulation program [1].

- Analysis of the results: The evaluation of physical properties is performed by taking time averages over the various movements of the system. Averages considered for a long time simulation correspond to averages of thermodynamic system.

Essentially, the fundamental difference between MD and MC is in the form of how to generate the trajectories of the system: while MD is done by the second law of Newton for calculating dynamic properties; in MC, probabilistic method is followed, thus excluding the possibility of calculating dynamic properties and only calculate static properties.

5.1.3 Collective Simulation

In order to simulate a physical system, it is necessary to clearly define the problem to be treated, that is, the type of properties to study, the range of parameters to be evaluated, and the required accuracy. Depending on the number of particles that use, the control variables, the interatomic potential, the type of average calculation, and simulation group (for produce microscopic to macroscopic properties system) in which the simulation will take place are decided. There are different types of collectives under which the simulations can be: collective canonical (NVT), collective isobaric–isothermal (NPT), large-canonical collective (μVT), collective micro-canonical (NVE), among others. Influences the choice of the group when calculating the fluctuations of the thermodynamic quantities, thereby enabling to calculate, for example, the heat capacity or the modulus of elasticity to name a few.

5.1.4 Interatomic Potential

The information on the energy potential between pairs of molecules that form a system is part of the fluid properties. Therefore, the quality of the simulation results depends on the model to describe the molecular interactions between the molecules of the fluid. A point of central importance both MD and MC is the choice of interatomic potential of the system to be simulated. The accuracy with which the interatomic potential represents actual interactions between particles depends on the quality of the results, because the more the details the potential has, the better the simulation.

However, the greater the potential complexity, the greater will also be the time required for computation. Undoubtedly, if what is sought is to test certain aspects of a model, it is best to use a simple potential that reproduces the essence of that model. On the other hand, if what you want is to simulate real materials, then the potential to contain the maximum information possible so as to reproduce the results not only qualitatively but also quantitatively.

For the construction of a potential, there are two essential steps: the first stage involves the development of potential model and the second stage is based primarily on redefining this potential again and again to find the optimal model. To create the potential is necessary to know the geometry of the molecule, the forces of repulsion and dispersion, and electrostatic effects involved in the system. In the molecular structure of a system, the length, bond angle, and torsion angle play an important role, since the position of each atom is determined by the nature of the chemical bonds with which their atoms will connect neighboring.

5.1.4.1 Forces of attraction–repulsion

The energy produced by the forces of attraction and repulsion between two particles is important to illustrate the energy in the system and is given by the following Equation 5.1:

$$U_{ij}(r) = \sum_{a}^{m} \sum_{b}^{n} 4\varepsilon_{abij} \left[\left(\frac{\sigma_{abij}}{r_{abij}} \right)^{12} - \left(\frac{\sigma_{abij}}{r_{abij}} \right)^{6} \right] \tag{5.1}$$

where is the distance between sites a and b, and y $\sigma_{abij} y \varepsilon_{abij}$ parameters are Lennard-Jones potential between a and b sites located in i and j molecules, as shown in Figure 5.2. The term r^{-12} describes the repulsion, and the term r^{-6} describes the attraction of the particles.

5.1.4.2 Electrostatic forces

The intermolecular forces between charged atoms or ions are undoubtedly the strongest physical interaction exists (Figure 5.3). The free energy between two charges q_{ai} and q_{bj} are characterized by Coulomb's Equation 5.2 whose interaction of electrostatic charges distributed in localized sites on the cation and anion is given by the following equation:

$$U_{ij}(r) = \sum_{a}^{m} \sum_{b}^{n} \frac{q_{ai} q_{bj}}{r_{abij}} \tag{5.2}$$

where q_{ai} is the loading a site to the i-th molecule, q_{bj} is the loading b site of the j-th molecule, and r_{abij} is the distance between these two charges.

5.1.5 Initial Conditions and Boundary Conditions

Specifying initial conditions for the position of each particle may be in a variety of ways, depending on the characteristics of the simulated system. Initially,

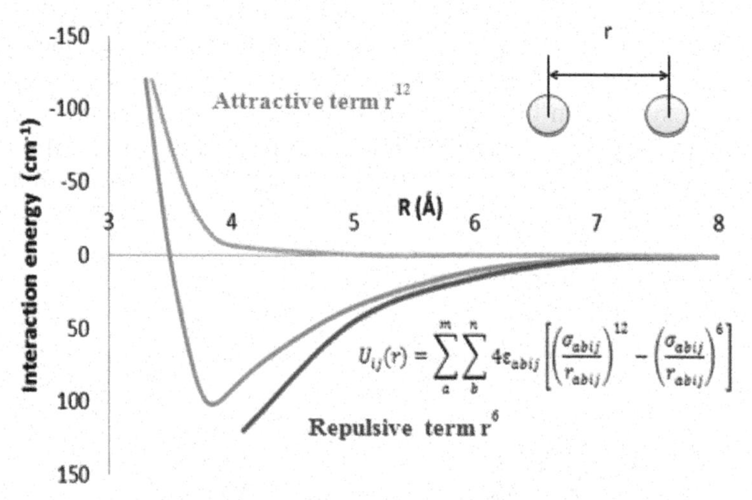

Figure 5.2 Graphical representation of Lennard-Jones potential [2].

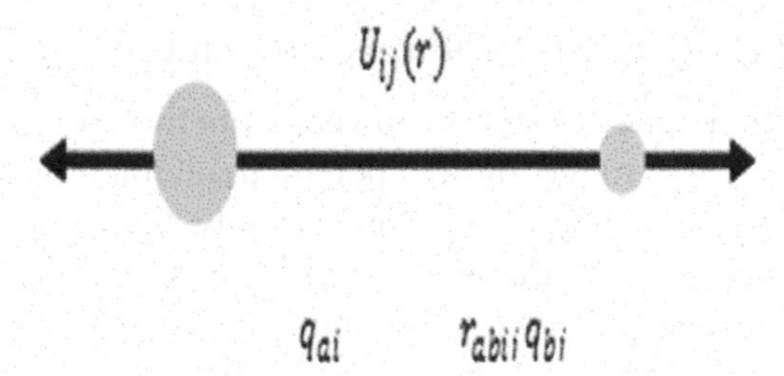

Figure 5.3 Force between two charges [2].

each particle is placed in positions that are assumed near-equilibrium in order to minimize the system energy. Another important aspect to be considered in the simulation is the right choice of boundary conditions or frequency of the system. Generally, when simulating a system, it is customary to use periodic boundary conditions (PBC) to avoid problems related to surface effects.

During the simulation, the molecules move within the original box and all periodic images move in the same way. When a molecule leaves the central box, one of the images comes from the opposite side. Thus, there are no walls in the central box avoiding surface effects, artificially imposing a system basis. The arrows in Figure 5.4 indicate that molecules cross and follow their path through the walls of simulation boxes in space replicated infinitely.

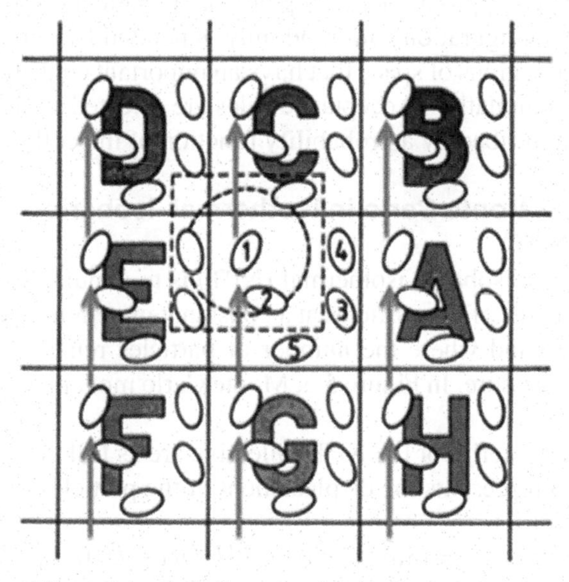

Figure 5.4 Periodic system in two dimensions [1].

5.1.6 Radio of Cutting and Condition of Minimum Image

The properties of the simulated systems can vary from those of a macroscopic system, depending on the range of the potential interaction between the molecules and the physical phenomenon under study. Generally, interaction potential is far reaching because the number of pair interactions grows as r^2, so that there are interactions between molecules and their images in other cases.

For this case, a cut is made in the simulation box in order to simulate a finite system. The condition can be represented minimum image from a particle i at which an imaginary box of the same size and shape is constructed that the simulation box, where the interaction of a particle i only does with regular image other particle j, and adds only on the particles within it as shown in Figure 5.4. The dotted box represents the condition of minimum image which is performed on the particle 1, while the dashed circle represents the radius cutting, where the particles contained within the circle (particles 2 and 4 and 5 imaginary) are taken into account the calculating interactions that have particle 1 in the simulation box.

5.1.7 Monte Carlo Simulation Technique

The MC method is a stochastic or probabilistic technique that uses a random number generator to approximate complex mathematical expressions. In MC, a number of configurations are randomly generated by certain rules of acceptance in those regions of space that have an important contribution in the average collective simulation. Acceptance rules are selected so that it occurs with a frequency described by a probability function of the collective.

5.1.7.1 Technical Monte Carlo in isothermal–isobaric group (NPT)

In MC, the collective isobaric–isothermal (NPT) is to simulate a box where there are displacements of the particles at a constant temperature and constant volume fluctuation, and where the number of particles remains unchanged during simulation pressure. In Figure 5.5, Monte Carlo movements are shown in a collective NPT.

The acceptance criteria for the NPT collective are as follows:

a) Equation 5.3 represents the displacement of the particles:

$$\mathrm{acc}\,(0 \rightarrow n) = \min(1, \exp\left[-\beta\left\{U(n) - U(0)\right\}\right] \qquad (5.3)$$

b) While the Equation 5.4 to volume change is given below:

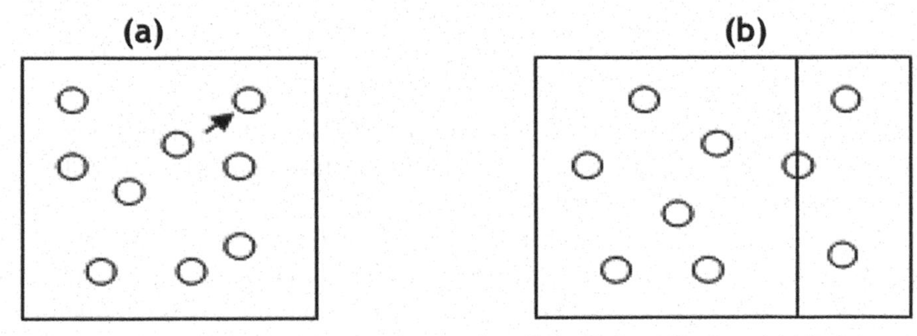

Figure 5.5 Movements Monte Carlo in isothermal–isobaric collective: (a) displacement (b) volume change [1].

$$\mathrm{acc}\,(0 \to n) = \min\left(1, \exp\left[-\beta\left\{U(n) - U(0) + P\,(V_n - V_0) - N\beta^{-1}\ln\left(\frac{V_n}{V_0}\right)\right\}\right]\right) \tag{5.4}$$

where 0 and *n* refer to the initial and new configurations, respectively, β is the temperature, U is the potential energy, *P* is the pressure, and *V* is the volume of the simulation box.

5.1.7.2 Insertion of test particle technique and Henry constant

When the solubility of a solute *i* in a solvent is very low, generally it is described by Henry's constant *hi* (see Equation 5.5), defined as follows:

$$h_i = \lim_{xi \to 0} \frac{f_i}{x_i} \tag{5.5}$$

where f_i y x_i are fugacity and mole fraction of solute i in the mixture. The relationship between the constant Henry's law and the excess chemical potential is given by Equation 5.6:

$$h_i = \frac{\rho_{\text{solvente}}}{\beta} \exp(\beta\mu_i^{\text{ex},\infty}) \tag{5.6}$$

where ρ_{solvente} is the density of the pure solvent at the temperature and pressure given, μ_i^{ex} is the chemical potential excess of solute, and $\mu_i^{\text{ex},\infty}$ is itself in the infinite dilution. A simple technique for calculating the chemical potential by MS was proposed by Widom in 1963. With this technique, the chemical potential of excess to a solute in a solvent at infinite dilution is estimated by Equation 5.7:

$$\mu_i^{\text{ex},\infty} = -\frac{1}{\beta}\ln\left[\frac{\langle V\exp\left(-\beta U_{i,\text{test}}\right)\rangle_{\text{NPT}}}{\langle V\rangle_{\text{NPT}}}\right] \tag{5.7}$$

where $<>_{\text{NPT}}$ is an average in the collective (NPT), V is the instantaneous volume, and $U_{i,\text{test}}$ is the energy due to the interaction of the test particle (solute) with all particles in the solvent.

In this method, at regular intervals during the simulation, a number of test particles insert with random position in a fluid solvent of N particles (Figure 5.6) are specified. Calculating $U_{i,\text{test}}$ of these particles it is obtained by averaging the Equation 5.7.

Substituting Equation 5.7 in 5.6, we have the Equation 5.8:

$$h_i = \frac{\rho_{\text{solvente}}}{\beta}\frac{\langle V\rangle_{\text{NPT}}}{\langle V\exp\left(-\beta U_{i,\text{test}}\right)\rangle_{\text{NPT}}} \tag{5.8}$$

where ρ_{solvent} is obtained of the solvent simulation in NPT collective.

5.1.8 Molecular System Description

The molecular description of a system is characterized in that an arbitrary system as if it were composed of individual entities each of which follows certain laws. The difference between the molecular descriptions AA and UA lies in the way in which molecular interactions are treated. In the case of molecular description, all atoms (AA) takes into account the interactions occurring in each of the atoms forming the molecule with the rest of the system. While in molecular description, united atoms (UA) only takes into account the interactions given by the union of atomic entities and not each of the atoms involved.

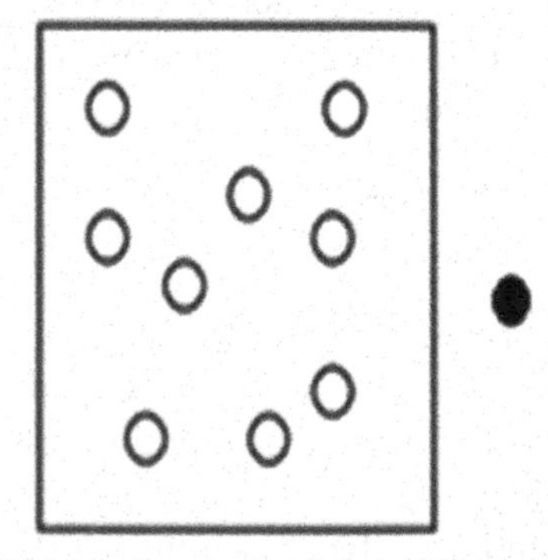

Figure 5.6 Inserting test particle for Widom method [1].

5.1.8.1 All atoms (AA)

The advantage of all atoms models is that they give a good account of molecular geometry and structure. The counterpart is that they require a great deal of computer time because of the large number of force centers, bending angles, and torsion angles involved. Also, the parameters assigned to a given atom depend on its environment (for instance, the properties assigned to hydrogen atoms bonded to oxygen atoms will differ from those assigned to hydrogen atoms bonded to nitrogen atoms).

5.1.8.2 United atoms (UA)

United atoms methods generally neglect the hydrogen atoms, while other atoms such as carbon, oxygen, and sulfur are represented by separate force centers. United atoms methods are often used for hydrocarbons because, with only one-third or one-quarter the number of force centers, they need less computer time. As computer time varies roughly with the square of the number of force centers, united atoms methods take about one-tenth the amount of computer time as all atoms methods. For fluid-phase equilibria and adsorption applications, united atoms methods are often used with success, especially for complex molecules which would be difficult to compute with all atoms models. However, the related simplification of molecular structure may be limiting for some applications.

5.1.9 Standard Monte Carlo Moves Involving a Single Box

5.1.9.1 Translation move

The first MC move that we will consider is *translation* of an individual molecule, through some acceptance criterion applied: if accepted, the new test configuration becomes the current configuration and all variables (energy, etc.) are updated (Figure 5.7b); if rejected, the old configuration remains the current configuration. This move does not change the internal conformation of the molecule.

In practice, the components of the translation vector are selected in a finite interval which is smaller than the simulation box.

5.1.9.2 Rotation move

A second type of move is *rotation* of an individual molecule through a random angle $\boldsymbol{a}$ in a randomly chosen direction. In this move (Figure 5.7c), the molecule is considered as a rigid body with internal bond distances, bending

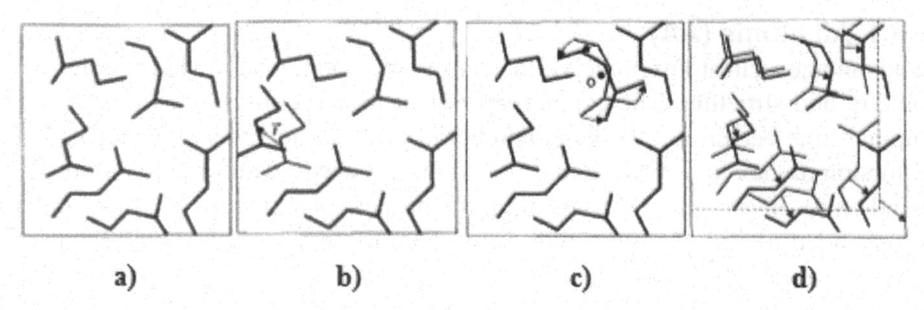

Figure 5.7 Elementary Monte Carlo moves of general purpose: (a) initial configuration, (b) translation of a single molecule, (c) rigid body rotation of a single molecule, and (d) volume change [3].

angles, and torsion angles preserved. In this move, the acceptance criterion is applied without any change.

5.1.9.3 Volume changes

When volume fluctuates, such as in the NPT collective, a specific MC move is used for *volume changes,* in which the simulation box is expanded. In this move (Figure 5.7d), the dimensionless positions of molecular centers of mass remain unchanged, as does the internal conformation of every molecule. Therefore, every molecule is translated, but the translation varies from one molecule to the other. The acceptance criterion incorporates the imposed pressure P.

5.1.9.4 Flip moves

In this move, a randomly chosen atom of a chain is rotated around the axis formed by its two immediate neighbors (Figure 5.8a), thereby preserving bond lengths. This move is particularly useful when it comes to relaxing the inner part of a long-chain molecule and flexible cyclic structures such as the cycloalkanes.

5.1.9.5 Reputation move

Another very efficient MC move to simulate long-chain molecules is *reputation* (Figure 5.8b). This involves suppressing a segment of one or more atoms at one end of a randomly selected molecule, and then adding an identical segment at the other end in a random position. Then, the acceptance criterion is applied to select or reject the move. In effect, the reverse of a flip move, reputation is applicable only to linear molecules.

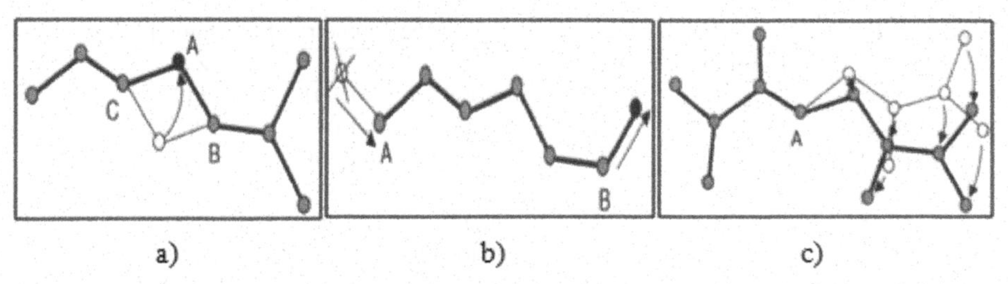

Figure 5.8 Elementary Monte Carlo moves contributing to the relaxation of flexible molecules: (a) flip, i.e., rotation of a single atom A around the axis B-C of its nearest neighbors, (b) reputation, (c) pivot, i.e., rotation of a part of a molecule around atom A [3].

5.1.9.6 Pivot move

When a molecule is composed of several more or less rigid parts (such as rings or branches) connected by a flexible chain, flip and reputation moves do not suffice to explore all possible internal configurations. Rotating part of the molecule around one of the atoms in a random rotation is the pivot move (Figure 5.8c), which then makes more adequate sampling possible. The acceptance probability applied is unchanged. This move is used for instance in the case of isoalkanes with multiple branches.

The following section describes the methodology to begin simulation IL System 1-n-butyl-3-methyl imidazolium hexafluorophosphate $[bmim]^+[PF6]^-$with molecular description AA and UA.

5.2 Methodology

5.2.1 Construction of the Cation and Anion

The geometry of the cation and anion to the molecular description of united atoms (UA) for thc IL was presented by Shah and Maginn [4] in 2004. Tables 5.1 and 5.2 show the initial coordinate system for this molecular description. Instead, the construction of the geometry description of all atoms (AA) is reported by Morrow and Maginn [5] and using the MCCCS Towhee 7.0.1 program [6].

5.2.2 Construction of the Simulation Box

Once realized the geometry of both molecular ions description of AA and UA, the simulation box is created. For the construction of the simulation box, 250 cations and anions 250 are used. In order to save simulation time, compared to

Table 5.1 xyz coordinates for the construction of the geometry of the anion [4]

	Position	Atom	*X*	*Y*	*Z*
ATOM	1	P	1.995	−1.001	0.512
ATOM	2	F1	1.428	−1.098	−1.066
ATOM	3	F2	0.502	−1.550	1.005
ATOM	4	F3	2.527	−2.527	0.414
ATOM	5	F4	2.462	−0.861	2.055
ATOM	6	F5	3.400	−0.413	−0.031
ATOM	7	F6	1.366	0.553	0.557

Table 5.2 xyz coordinates for construction of the molecular geometry of the cation with description of atoms bonded (UA)[4]

	Position	Atom	*X*	*Y*	*Z*
ATOM	1	N1	−2.0478	0.682	0.272
ATOM	2	C2	−1.322	−0.100	−0.530
ATOM	3	N3	−2.041	−1.177	−0.852
ATOM	4	C4	−3.274	−1.083	−0.232
ATOM	5	C5	−3.279	0.082	0.469
ATOM	6	C6	−1.542	−2.310	−1.640
ATOM	7	C7	−1.574	1.940	0.888
ATOM	8	C8	−1.993	3.187	0.103
ATOM	9	C9	−1.340	3.348	−1.278
ATOM	10	C10	0.173	3.582	−1.238

a system positioned neatly particles, first a parcel called Packmol is used [7] to create a simulation box with disordered system-related particles in which particles of the system are mixed and to make interactions between cations and anions. The simulation box is only necessary to insert coordinates of a particle, the number of particles of each type, and the space limitations that have between them. Then, you need to create an input file of MCCCS Towhee (Table 5.3) for molecular description of the system.

5.2.3 System Simulation Parameters

The MC simulation technique in the isobaric–isothermal collective NPT to calculate the thermodynamic properties was used. In order to avoid problems of surface effects, CBP is used. To calculate the intermolecular interactions of the system, the Lennard-Jones (LJ) interactions were used for molecular descriptions AA and UA because they are highly related to attraction–repulsion forces between particles (Table 5.3). Also, interactions between two different

Table 5.3 Input file MCCCS Towhee for the determination of thermodynamic properties with molecular description of all atoms (AA) [6]

Input Format	'Towhee'	Ensemble	'npt'
Temperature	298.0	Pressure	101.3
Nmolty	2	Nmolectyp	250 250
Numboxes	1	Stepstyle	'cycles'
Nstep	1000	Controlstyle	'equilibration'
Potentialstyle	'internal'	Ffnumber	1
ff_filename	/towheebase/Force Fields/towhee_ff_ Morrow2002	classical_potential	'Lennard-Jones'
classical_mixrule	'Lorentz-Berthelot'	Lshift	.false.
Ltailc	.true.	rmin	0.09d0
rcut	20.0d0	rcutin	18.0d0
electrostatic_form	'coulomb'	Coulombstyle	'ewald_fixed_kmax'
Kalp	5.6d0	Kmax	5
Dialect	1.0d0	linit	.true.
Initboxtype	'dimensions'	Initstyle	'coords' 'coords'
Initlattice	'none' 'none'	Initmol	250 250
inix iniy iniz	8 8 8	Hmatrix	43.0 0.0 0.0
0.0 43.0 0.0	0.0 0.0 43.0	Pmvol	0.003
Pmvlpr	1.0	Rmvol	0.5
Tavol	0.2	Pmpivot	0.1d0
Pmpivmt	1.0d0 0.0d0	pmtracm	0.2d0
Pmtcmt	0.5d0 1.0d0	Rmtrac	5.0d0 0.9d0
tatrac	0.5d0	Pmrotate	1.0d0
pmromt	0.5d0 1.0d0	Rmrot	0.5d0 0.5d0
Tarot	0.5d0	#mim CATION	input_style
'basic connectivity map'	Nunit	25	Nmaxcbmc
25	Lpdbnames	F	Forcefield
'Morrow2002'	charge_assignment	'bond increment'	unit ntype
1 'N1'	Vibration	3	2 5 7
improper torsion	0	unit ntype	2 'C2'
Vibration	3	1 3 11	improper torsion
0	unit ntype	3 'N3'	Vibration
3	2 4 6	improper torsion	0
unit ntype	4 'C4'	Vibration	3
3 5 12	improper torsion	0	unit ntype
5 'C5'	Vibration	3	1 4 13
improper torsion	0	unit ntype	6 'C6'
Vibration	4	3 14 15 16	improper torsion
0	unit ntype	7 'C7'	Vibration
4	1 8 17 18	improper torsion	0

(*Continued*)

Table 5.3 Continued

Input Format	'Towhee'	Ensemble	'npt'
unit ntype	8 'C8'	Vibration	4
7 9 19 20	improper torsion	0	unit ntype
9 'C9'	Vibration	4	8 10 21 22
improper torsion	0	unit ntype	10 'C10'
vibration	4	9 23 24 25	improper torsion
0	unit ntype	11 'H1'	vibration
1	2	improper torsion	0
unit ntype	12 'H2'	Vibration	1
4	improper torsion	0	unit ntype
13 'H3'	Vibration	1	5
improper torsion	0	unit ntype	14 'H4'
vibration	1	6	improper torsion
0	unit ntype	15 'H5'	vibration
1	6	improper torsion	0
unit ntype	16 'H6'	Vibration	1
6	improper torsion	0	unit ntype
17 'H7'	Vibration	1	7
improper torsion	0	unit ntype	18 'H8'
vibration	1	7	improper torsion
0	unit ntype	19 'H9'	vibration
1	8	improper torsion	0
unit ntype	20 'H10'	Vibration	1
8	improper torsion	0	unit ntype
21 'H11'	Vibration	1	9
improper torsion	0	unit ntype	22 'H12'
vibration	1	9	improper torsion
0	unit ntype	23 'H13'	Vibration
1	10	improper torsion	0
unit ntype	24 'H14'	Vibration	1
10	improper torsion	0	unit ntype
25 'H15'	Vibration	1	10
improper torsion	0	#PF6 ION	input_style
'basic connectivity map'	Nunit	7	Nmaxcbmc
7	Lpdbnames	F	Forcefield
'Morrow2002'	charge_assignment	'bond increment'	unit ntype
1 'P'	Vibration	6	2 3 4 5 6 7
improper torsion	0	unit ntype	2 'F'
vibration	1	1	improper torsion
0	unit ntype	3 'F'	Vibration
1	1	improper torsion	0
unit ntype	4 'F'	Vibration	1
1	improper torsion	0	unit ntype

Input Format	'Towhee'	Ensemble	'npt'
5 'F'	Vibration	1	1
improper torsion	0	unit ntype	6 'F'
Vibration	1	1	improper torsion
0	unit ntype	7 'F'	Vibration
1	1	improper torsion	0

sites of LJ atoms were determined using the rules of combination of Lorentz-Berthelot [2], and force fields of Maginn and Morrow 2002 [5], and Shah and Maginn 2004 [4] were used in the simulation. These fields are used to describe the ion for molecular descriptions AA and UA (Tables 5.4–5.8).

5.2.4 Calculation of Thermodynamic Properties

To evaluate the physical properties with time averages, information acquired on the different configurations of the system are taken movements. Mean values were obtained for a number of cycles sufficiently long. How soon were determined thermodynamic and structural properties of the simulated MC

Table 5.4 Parameters used for the potential $[bmim]^{+}$ $[PF6]^{-}$ for molecular description UA. Values of sigma, epsilon, and charge of each site [4]

Atom	σii (Å)	ε (KJ/mol)	q (e)
N1	3.250	0.8539	0.111
N3	3.250	0.8539	0.133
C2	3.880	0.5328	0.233
C4	3.880	0.5328	0.040
C5	3.880	0.5328	–0.010
C6	3.775	1.0403	0.183
C7	3.905	0.5929	0.195
C8	3.905	0.5929	–0.066
C9	3.905	0.5929	0.128
C10	3.905	0.8803	–0.043
Charge of the Cation			0.904
P	3.740	1.0054	1.460
F1	3.118	0.3066	–0.394
F2	3.118	0.3066	–0.394
F3	3.118	0.3066	–0.394
F4	3.118	0.3066	–0.394
F5	3.118	0.3066	–0.394
F6	3.118	0.3066	–0.394
Charge of the Anion			0.904

Table 5.5 Parameters of torsion of dihedral angles for UA molecular description [5]

Dihedral Angle	V0 KJ/mol	V1 KJ/mol	V2 KJ/mol	V3 KJ/mol
$C_7-C_8-C_9-C_{10}$	0.00	5.89	−1.13	13.17[a]
$N_1-C_7-C_8-C_9$	0.00	11.41	0.96	2.05[b]
$C_2-N_1-C_7-C_8$	0.00	−5.85	−1.80	0.00[c]
$C_5-N_1-C_7-C_8$	0.00	−5.85	−1.80	0.00[c]

[a]OPLS-UA potential parameters for the n-alkane.
[b]OPLS-AA potential parameters for the n-methyl formamide.
[c]OPLS-UA potential parameters for the ammonium ion.

system is described [10]. In MC, the thermodynamic properties such as *Cp*, *Cv*, enthalpy, chemical potential of the solute, thermal expansion coefficient, isothermal compressibility, Joule–Thomson coefficient, and sound velocity are described below by the following equations.

5.2.4.1 Thermal expansion coefficient (α_P)

The thermal expansion coefficient can be estimated by running a series of simulations with constant pressure and variable temperatures and by controlling fluctuations in the volume and enthalpy produced during a simulation NPT. Formally, this is written as follows [4]:

$$\alpha_P = \frac{1}{< V > k_B T^2} (< VH > - < V >< H >) \tag{5.9}$$

where k_B is the Boltzmann constant, H is the enthalpy, and $<>$ denotes the average of the group simulation.

5.2.4.2 Isothermal compressibility coefficient (k_T)

In a simulation, the isothermal compressibility can be done by volume fluctuations in the system controlled for constant pressure at a given temperature, and then k_T remains as in Equations 5.10 and 5.11 [4]:

$$k_T = \frac{1}{V}\left(\frac{\partial V}{\partial P}\right)_T \tag{5.10}$$

$$k_T = \frac{< V^2 > - < V >^2}{k_B T < V >} \tag{5.11}$$

where k_B is the Boltzmann constant.

Table 5.6 Parameters used for the potential $[bmim]^+$ $[PF6]^-$ for molecular description AA. Values of sigma, epsilon, and charge of each site [4]

Atom	σii (Å)	ε (KJ/mol)	q (e)
N1	3.296	1.0066	0.111
N3	3.296	1.0066	0.133
C2	3.207	0.2513	0.056
C4	3.207	0.2513	–0.141
C5	3.207	0.2513	–0.217
C6	4.053	0.1010	–0.157
C7	4.053	0.1010	0.095
C8	3.581	0.2814	–0.122
C9	3.581	0.2814	0.256
C10	3.634	0.3920	–0.209
H1	1.603	0.2309	0.177
H2	2.615	0.0396	0.181
H3	2.615	0.0396	0.207
H4	2.351	0.1106	0.125
H5	2.351	0.1106	0.073
H6	2.351	0.1106	0.142
H7	2.387	0.1106	0.055
H8	2.387	0.1106	0.045
H9	2.387	0.1407	0.055
H10	2.387	0.1407	0.001
H11	2.387	0.1407	–0.029
H12	2.387	0.1407	–0.099
H13	2.387	0.1202	0.051
H14	2.387	0.1202	0.040
H15	2.387	0.1202	0.075
Charge of the Cation			0.904
P	3.830	2.9442	1.457
F	3.029	0.4534	–0.3935
F	3.029	0.4534	–0.3935
F	3.029	0.4534	–0.3935
F	3.029	0.4534	–0.3935
F	3.029	0.4534	–0.3935
F	3.029	0.4534	–0.3935
Charge of the Anion			–0.904

5.2.4.3 Isochoric and isobaric heat capacity (C_v and C_p)

Indeed, the heat capacity of a system can be any value from minus infinity to plus infinity. However, only values are of fundamental importance.

As they are not equal, it is important to find a relationship between them. Involving the relationship is as follows (Equations 5.12 and 5.13) [11]:

Table 5.7 Parameters of torsion of dihedral angles for AA molecular description [4]

Dihedral Angle	V0 KJ/mol	V1 KJ/mol	V2 KJ/mol	V3 KJ/mol
$C_7-C_8-C_9-C_{10}$	0.00	5.89	−1.13	13.17[a]
$N_1-C_7-C_8-C_9$	0.00	11.41	0.96	2.05[b]
$C_2-N_1-C_7-C_8$	0.00	−5.85	−1.80	0.00[c]
$C_5-N_1-C_7-C_8$	0.00	−5.85	−1.80	0.00[c]

[a]OPLS-UA potential parameters for the n-alkane.
[b]OPLS-AA potential parameters for the n-methyl formamide.
[c]OPLS-UA potential parameters for the ammonium ion.

Table 5.8 Parameters for the potential of CO_2. Values of sigma, epsilon, and charge of each site [8, 9]

Atom	σii (Å)	ε (KJ/mol)	q (e)
C	2.800	0.270	0.700
O	3.050	0.790	−0.350
O	3.050	0.790	−0.350
Charge of the CO_2			0.000

$$C_p - C_v = P\left(\frac{\partial V}{\partial T}\right)_P \tag{5.12}$$

Thus, clearing C_v is given by Equation 5.13:

$$C_v = C_p - \left(T < V > \frac{\alpha_{\mathrm{P}}^2}{k_{\mathrm{T}}}\right) \tag{5.13}$$

where V is the volume, T is the temperature, α_{P} is the coefficient of thermal expansion, and k_{T} is the coefficient of isothermal compressibility [12].

5.2.4.4 Joule–Thomson coefficient (μ_{JT})

The Joule–Thomson coefficient (Equation 5.14) is a ratio of experimentally determined limiting change at constant enthalpy.

$$\mu_{\mathrm{JT}} = \lim_{\Delta P \to 0}\left(\frac{\Delta T}{\Delta P}\right)_H = \left(\frac{\partial T}{\partial P}\right)_H \tag{5.14}$$

Therefore, the Joule–Thomson coefficient is given by Equation 5.15 [13]:

$$\mu_{\mathrm{JT}} = \frac{< V >}{C_{\mathrm{p}}}\left[T\alpha_{\mathrm{P}} - 1\right] \tag{5.15}$$

where V is the volume, T is the temperature, C_{p} is the isobaric heat capacity, and α_{P} is the thermal expansion coefficient. The Joule–Thomson effect is used for gas liquefaction [14].

5.2.4.5 Speed of sound (*u*)

The sound velocity measures the speed that has a wavelength in the middle where it spreads; that is to say, how fast a wave would travel in the ionic liquid (Equation 5.16). This velocity is given by the following equation [12]:

$$\sqrt[2]{u^2} = \sqrt[2]{-\frac{C_\mathrm{P}}{C_\mathrm{v}}\frac{MN_\mathrm{A}}{N}\frac{<V>}{k_\mathrm{T}}} \tag{5.16}$$

where $C_\mathrm{p} y C_\mathrm{v}$ are the heat capacities, M is the molecular mass, N is the number of particles, N_A is the Avogadro number, V is the volume, and k_T is the isothermal compressibility.

5.2.4.6 Chemical potential of the solute (μ_2^{ex})

The chemical potential is usually determined using the approximation of Widom, in which a test particle is inserted into the system and the resulting change in the potential energy is calculated. The insertion of this particle causes, and potential energy change is given by Equation 5.17 [15]:

$$\mu_2^{\mathrm{ex}} = -k_\mathrm{B}T \ln \frac{< V_{\mathrm{exp}}(-\beta U_\mathrm{g}) >}{< V >} \tag{5.17}$$

where V is the volume, T is the temperature, k_B is the Boltzmann constant, U_g is the interaction energy of the solute molecules in the solvent, and the parameter $\beta = \left(\frac{1}{k_\mathrm{B}T}\right)$ [4].

5.2.4.7 Henry constant (*h*)

The solubility of a dilution of solute 2 in a solvent 1 is generally expressed in terms of the law of Henry. In a simulation, Henry's constant can be calculated with the Equation 5.18:

$$h_{2,1} = k_\mathrm{B}T\rho_1 \exp(-\beta\mu_2^{\mathrm{ex}}) \tag{5.18}$$

where ρ_1 is the liquid density, T is the temperature, k_B is the Boltzmann constant, μ_2^{ex} is the chemical potential, and the parameter $\beta = \left(\frac{1}{k_\mathrm{B}T}\right)$ [4].

5.2.5 Calculation of Structural Properties

The structural properties (Equation 5.19) can be described through the pair distribution function *g*(*r*) and the static structure factor *S*(*q*). In the case of liquids and amorphous materials, this module depends on only |q|:

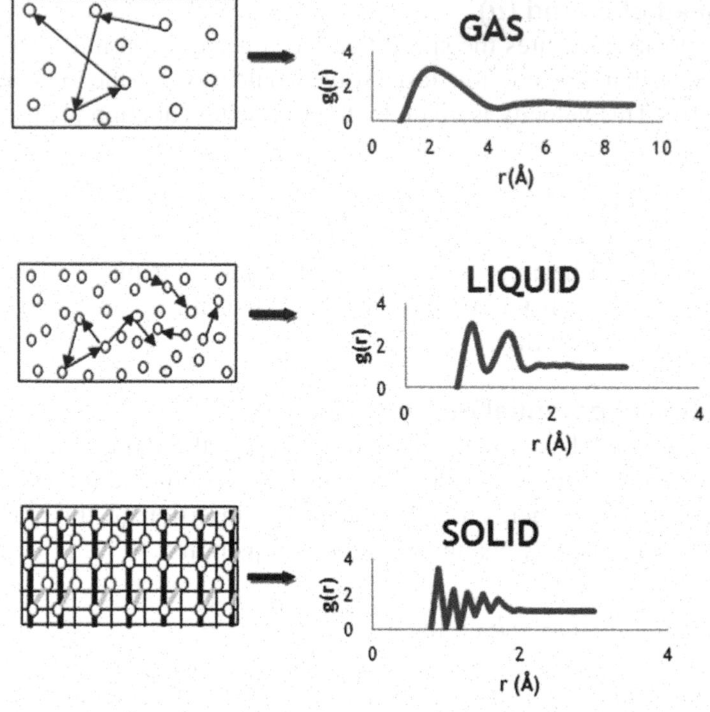

Figure 5.9 Radial distribution function (FDR).

$$S(q) = 1 + 4\pi\rho \int_0^R r^2 \left[g\left(r\right) - 1\right] \frac{\mathrm{sen}(qr)}{qr} dr \tag{5.19}$$

where the value of R must be chosen smaller than half the length of the simulation box. Figure 5.9 shows schematically the typical shape having the pair distribution function $g(r)$ and the diffraction pattern for gases, liquids, and solids referring to the arrangement of their atoms in real space.

5.3 Results and Discussions

5.3.1 Molecule Construction

The *xyz* coordinates for the construction of the geometry of cation and anion with molecular description of AA and UA were obtained from Tables 5.1 and 5.2 and 7.0.1 MCCCS Towhee software [6]. Figures 5.10–5.12 show the geometries for both ions of LI; therefore, the cation and the anion are used. Figure 5.10 shows atomic representation of cation $[\text{bmim}]^+$ with the molecular

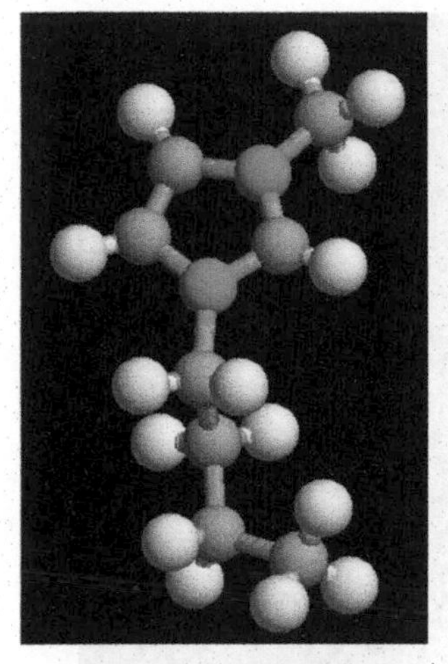

Figure 5.10 Atomic representation of the cation $[bmim]^+$. All atoms (AA) molecular description.

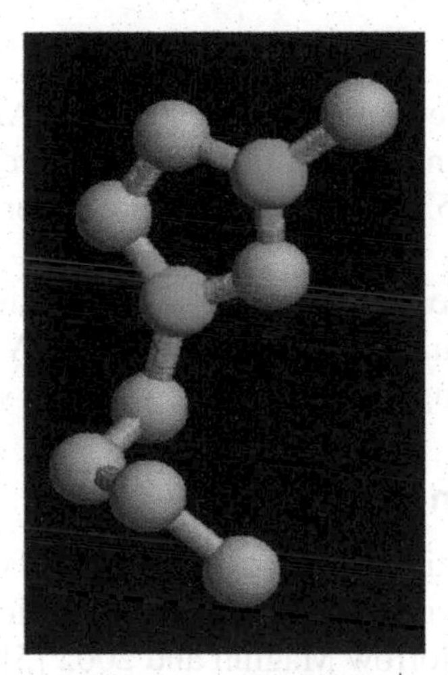

Figure 5.11 Atomic representation of the cation $[bmim]^+$. United atoms (UA) molecular description.

description of AA. As can be seen in Figure 5.10, the cation is represented as follows: nitrogens are represented by purple spheres, carbon by gray spheres, and hydrogen atoms by white spheres. In Figure 5.11, the hydrogen atoms are implied by the gray areas represented by the carbon. In Figure 5.12, the central sphere is represented by the phosphorus atom and around it are fluorine atoms.

Figure 5.12 Atomic representation of the anion $[PF_6]^-$.

5.3.2 Simulation Box

Once constructed, the geometry of molecules UA and AA proceeded to build the simulation box. Two hundred and fifty ion pairs were used to create the simulation box. In Figures 5.13 and 5.14, MS boxes for both the description of AA to UA are presented.

For the simulation box for AA, greater conglomeration of atoms is shown than UA. This is because in the description of the AA molecular hydrogen atoms, explicit cation is being observed and denser the simulation box.

5.3.3 Data Entry System

The input data for the simulation is shown in Table 5.9 in which you specify the following: the type of collective (NPT), system conditions, force field (Shah and Maginn 2004 [4] and Morrow Maginn and 2002 [5]), likely to rotational movements, volume, and geometry of the molecule.

5.3.4 Equilibration Phase of System

At this stage, the system density was monitored as a reference parameter for determining the IL properties. The objective of this phase equilibration was first to validate the method and then to determine the system properties. Table 5.10 shows the change of thermodynamic properties at different cycles

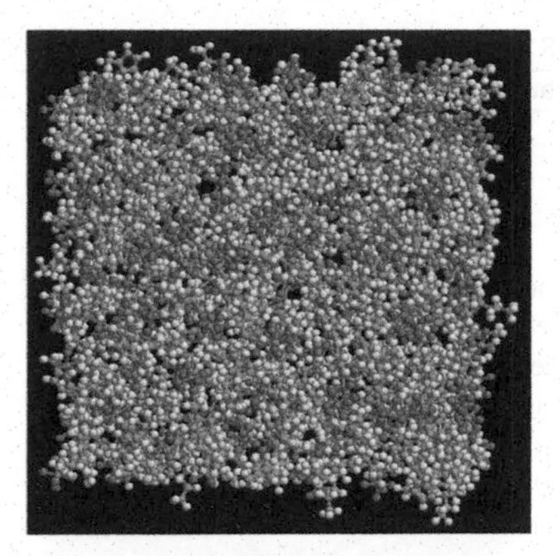

Figure 5.13 Simulation Box AA.

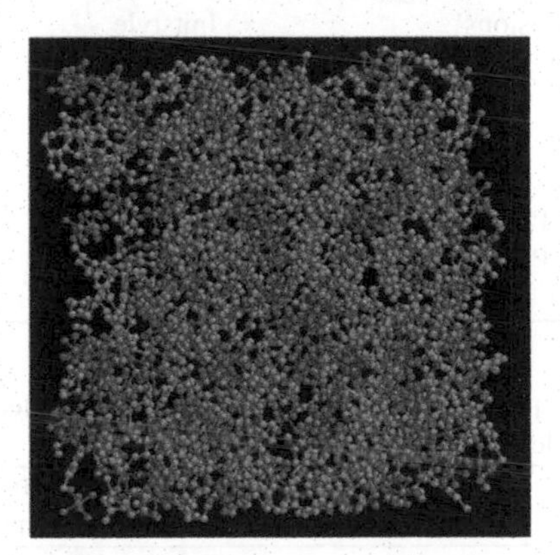

Figure 5.14 Simulation Box UA.

MC during the equilibration for the description of AA, while in Table 5.11, it is shown for UA. The purpose of equilibrium stage is to bring the system to a point where the values obtained do not show large fluctuations.

The experimental data reported for this system are taken from literature Shah and Maginn [8] with the following values: density 1.3603 gr/cm^3, pressure 101.3 kPa, Henry constant 53.4 $\pm$ 0.03 bar, constant thermal expansion 0.000611°K–1, isothermal compressibility constant 41.95E-11 Pa^1, enthalpy –16.1$\pm$2.2 kJ/mol°K, and entropy –53.2 $\pm$ 6.9 J/mol°K, respectively. In the case of stage equilibration, it is observed that a gradual increase in the density value as time goes by MC cycles up to a value ranging from 1.29 to 1.30 gr/cm^3 in molecular description AA, while UA arrives an oscillating value between 1.26 and 1.27 gr/cm^3 as shown in Tables 5.10 and 5.11.

Table 5.9 Input File MCCCS Towhee for the determination of thermodynamic properties [6]

Input Format	'Towhee'	Ensemble	'npt'
Temperature	298.0	Pressure	101.3
Nmolty	2	Nmolectyp	250 250
Numboxes	1	Stepstyle	'cycles'
Nstep	1000	Controlstyle	'equilibration'
Potentialstyle	'internal'	Ffnumber	1
ff_filename	/towheebase/ForceFields/ towhee_ff_Morrow2002	classical_pot	'Lennard-Jones'
classical_mixrule	'Lorentz-Berthelot'	Lshift	.false.
Ltailc	.true.	rmin	0.09d0
rcut	20.0d0	rcutin	18.0d0
electrostatic_form	'coulomb'	Coulombstyle	'ewald_fixed_km'
Kalp	5.6d0	Kmax	5
Dialect	1.0d0	linit	.true.
Initboxtype	'dimensions'	Initstyle	'coords' 'coords'
Initlattice	'none' 'none'	Initmol	250 250
inix iniy iniz	8 8 8	Hmatrix	43.0 0.0 0.0 0.0 43.0 0.0 0.0 0.0 43.0
Pmvlpr	1.0 Pmvol 0.003	pmtracm	0.2d0
Tavol	0.2 Pmpivot 0.1d0	Pmrotate	1.0d0
Pmrotate	1.0d0 Rmrot 0.5d0 0.5d0	Tarot	0.5d0

Table 5.10 Thermodynamic properties at different cycles MC in collective NPT during the equilibration phase of AA molecular description

MC Cycles	Density (gr/cm^3)	Pressure (kPa)	Energy (kJ/mol)	Cp (J/mol K)	Enthalpy (kJ/mol K)	Entropy (kJ/mol K)	Time of Simulation (h)
0	1.2847	261.2E5	–	–	–	–	0
8000	1.2227	–545.5E0	133.8E4	17.55E2	44.9E2	44.89E5	158.5
17000	1.2831	659.6E0	133.6E4	19.09E2	44.8E2	44.83E5	884.1
20000	1.2923	–175.9E1	133.5E4	23.85E2	44.8E2	44.81E5	1124.5
24000	1.3009	–173.0E0	133.5E4	17.49E2	44.8E2	44.80E5	1478

–Data not determined.

5.3.5 Production Phase of System

Once balanced the system, the thermodynamic properties of the system are determined. Density, energy, heat capacity, enthalpy, entropy, chemical potential, thermal expansion coefficient, isothermal compressibility, Joule–Thomson coefficient, sound velocity, the radial distribution functions with phosphorus anion of the imidazolium ring carbons cation, and Henry constant were obtained. For determining the Henry constant, force field Trappe-EH [16]

Table 5.11 Thermodynamic properties at different cycles MC in collective NPT during the equilibration phase of UA molecular description

MC Cycles	Density (gr/cm^3)	Pressure (kPa)	Energy (kJ/mol)	Cp (J/mol K)	Enthalpy (kJ/mol K)	Entropy (kJ/mol K)	Time of Simulation (hrs)
0	0.9675	−797.8E1	–	–	–	–	0
19000	1.2294	185.4E1	−89.64E3	12.75E2	−30.0E1	−30.04E4	432
28000	1.2523	334.1E1	−90.16E3	11.36E2	−30.2E1	−30.19E4	695
38000	1.2616	−179.1E1	−90.42E3	13.54E2	−30.4E1	−30.38E4	1039
48000	1.2699	−199.8E1	−90.75E3	20.76E2	−30.5E1	−30.49E4	1417

–Data not determined.

to the molecule of carbon dioxide was used. Tables 5.12 and 5.13 show the thermodynamic properties during the production phase for the description of AA and UA.

The values obtained for the density, energy, enthalpy, and entropy fluctuate as in the equilibration stage from 1.29 to 1.30 gr/cm^3 for density in AA and 1.27 gr/cm^3 for UA. For entropy values obtained range from 44.78E2 to 44.80E2 KJ/mol°K for AA and −30.73E1 to −30.58E1 KJ/mol°K for UA. Energy and enthalpy values are maintained as in the equilibration stage of AA and UA. The Cp and pressure continued to have ups and downs in the values obtained during this stage compared with equilibration but more moderately. Similarly, the chemical potential, Henry constant, thermal expansion coefficient, isothermal compressibility, Joule–Thomson coefficient, and sound velocity up and down along the production phase.

Table 5.12 Thermodynamic properties at different cycles MC in collective NPT during the production phase of AA molecular description

MC Cycles	Density (gr/cm^3)	Pressure (kPa)	Energy (kJ/mol)	Cp (J/mol K)	Enthalpy (kJ/mol K)	Entropy (kJ/mol K)	Time Simulation (h)
24100	1.3047	−417.4E1	133.5E4	15.13E2	44.8E2	44.78E2	1562
26100	1.3055	256.5E1	133.5E4	11.95E2	44.8E2	44.80E2	1860.5
28100	1.2996	−138.0E2	133.5E4	10.80E2	44.8E2	44.77E2	2149
29100	1.3058	−520.4E1	133.5E4	24.72E2	44.8E2	44.78E2	2277

MC Cycles	μ (J/mol)	H(bar)	α_P(K^{-1})	k_T(Pa^{-1})	μ_{JT}(K/MPa)	u(m/s)	Time Simulation
24100	−8.34E3	7.8675	0.00085	2.9898E-11	−4.45E-23	313.0949	1562
26100	−8.85E3	6.8660	0.0037	1.2720E-10	8.55E-24	63.6466	1860.5
28100	−10.52E3	3.2478	−0.00062	9.8100E-12	−9.96E-23	364.1709	2149
29100	−8.701E3	6.7937	0.000598	7.2400E-11	−3.00E-23	580.7447	2227

Table 5.13 Thermodynamic properties at different cycles MC in collective NPT during the production phase of UA molecular description

MC Cycles	Density (g/cm^3)	Pressure (kPa)	Energy (kJ/mol)	Cp (J/mol K)	Enthalpy (kJ/mol K)	Entropy (kJ/mol K)	Time Simulation (hrs)
49000	1.2710	–730.7E1	–90.73E2	12.10E1	–30.6E1	–30.58E1	1453
50000	1.2734	–729.7E0	–90.84E2	11.03E2	–30.5E1	–30.50E1	1488.8
51000	1.2755	–468.6E1	–90.93E2	12.09E2	–30.6E1	–30.60E1	1524.5
52000	1.2765	–101.4E2	–91.02E2	13.34E2	–30.7E1	–30.73E1	1719
MC Cycles	μ (J/mol)	h (bar)	$_P(K^{--1})$	$k_T(Pa^{--1})$	$_{JT}$(K/MPa)	U(m/s)	Time Simulation
49000	–7.970E3	8.8884	0.000013	9.98E-12	–1.13E-22	N/A	1453
50000	–7.035E3	12.98	–0.00031	1.13E-11	–9.17E-23	771.9083	1488.8
51000	–8.062E3	8.5912	–0.00063	2.38E-11	–9.09E-23	387.0124	1524.5
52000	–6.798E3	14.32	–0.00025	1.01E-11	–7.45E-23	1071.815	1719

During the simulation system, a big difference was observed between the values obtained for the internal energy, enthalpy, and entropy between AA force field and those obtained by UA. This indicates that the presence of the hydrogen atoms of the force field AA has a high inference on thermodynamic properties obtained because the hydrogen atoms are not considered as a union of atoms that can move, flex, and torqued more easily that if attached, slightly modifying the properties of the system.

In the insertion method, for determining particle, Henry's constant –was conducted under two methods: explicit and Widom [17] methods. Table 5.14 shows the difference between the methods used to calculate Henry's constant.

Widom's method for inserting the test particle yields better results than using explicit AA in some cases, while in the description of UA, both methods yield the same results. This is because the molecular description of AA is much

Table 5.14 Difference between the methods used to calculate the constant of Henry (explicit and Widom) for molecular description of AA and UA

MC Cycles.	AA		MC Cycles.	UA	
Production Phase	Explicit	Widom	Production Phase	Widom	Explicit
24100	7.8675	–	49000	8.8842	8.8842
25100	5.2668	–	50000	12.98	12.98
26100	6.8660	6.4012	51000	8.5912	8.5912
27100	3.2478	7.1291	56000	14.32	14.32
28100	–	3.2478	–	–	–
29100	–	6.7937	–	–	–

–Data not determined.

more accurate than UA since it involves all the interactions of all the atoms, while UA only considers the interactions between the particles together taking great inference calculating solubility CO_2-LI. In order to better appreciate the data presented in the above tables, Figures 5.15 and 5.16 exhibit graphically the behavior of the density and pressure system along the equilibration and production phases.

During the equilibration, the density of IL showed an increase up to a value of 1.30 gr/cm^3 for AA and 1.27 gr/cm^3 for UA; after this value, the density is kept constant with slight fluctuations (Figure 5.15), indicating that the system had reached equilibrium. In the case of pressure, however (Figure 5.16), deviations were very high considerably affecting the values obtained for the Henry constant, thermal expansion coefficient, Joule–Thomson coefficient, sound velocity, chemical potential, *C*p, enthalpy, and entropy because they are closely related to the system pressure.

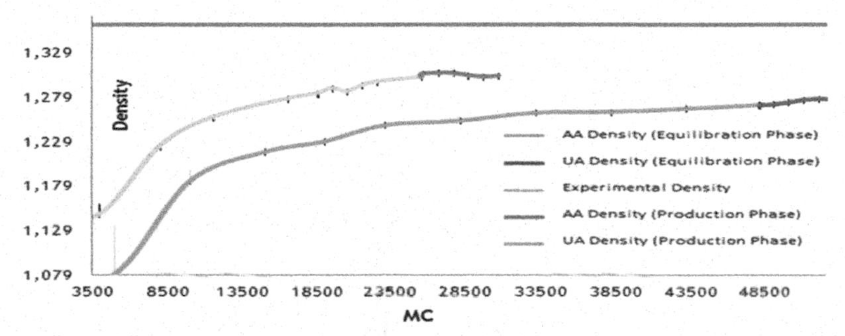

Figure 5.15 Monitoring the density value of ionic liquid 1-N-butyl-3-methyl imidazolium hexafluorophosphate and typical deviation along equilibration and production phases to different Monte Carlo (MC) cycles.

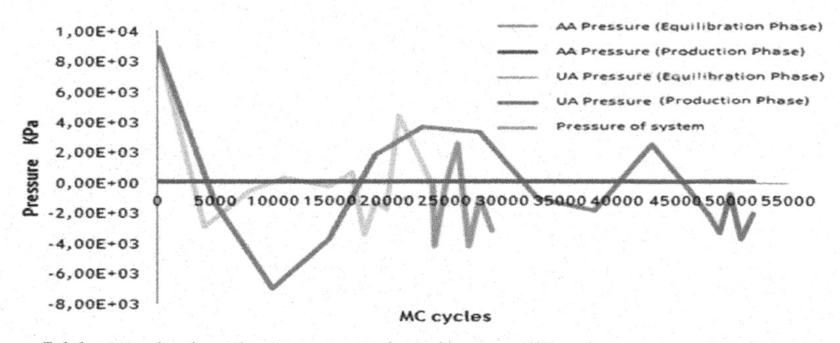

Figure 5.16 Monitoring the pressure value along equilibration and production phases to different Monte Carlo (MC) cycles.

5.3.6 Radial Distribution Functions of Ionic Liquid

Better information on the structure of fluid can be extracted by specific consideration of the sites of the ions. It is particularly convenient to investigate the FDR between the phosphorus atom (P) and the carbon atoms of the imidazolium ring. Figures 5.17 and 5.18 show the numbering scheme of the cation for the description of AA and UA, while for anion, scheme is presented in Figure 5.19.

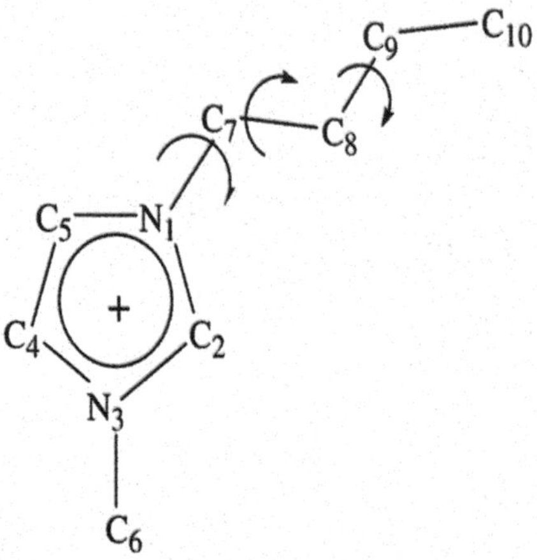

Figure 5.17 Outline numbering cation $[bmim]^+$ with UA [18].

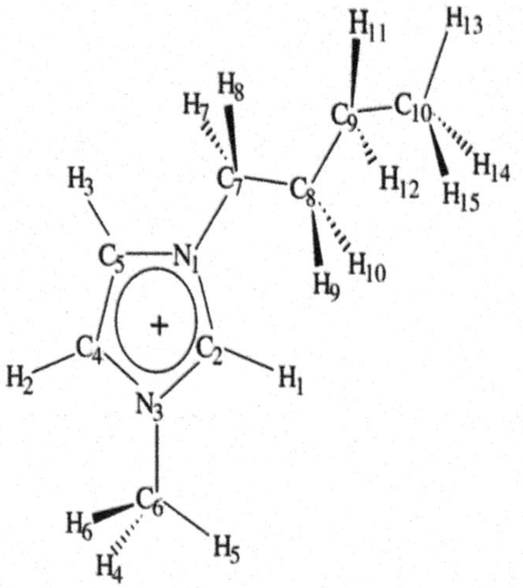

Figure 5.18 Outline numbering cation $[bmim]^+$ with AA [18].

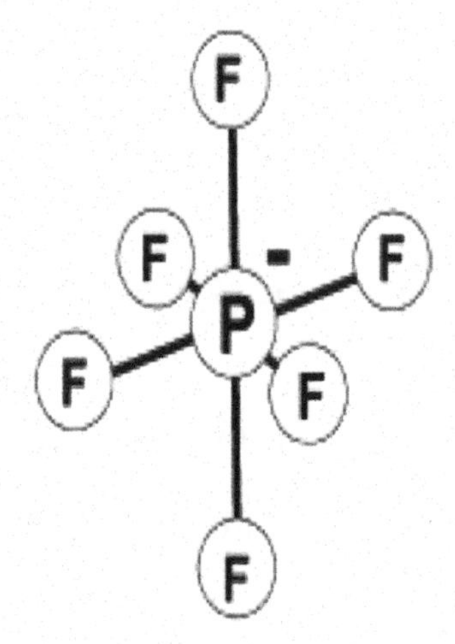

Figure 5.19 Representation of anion $[PF6]^-$ [14].

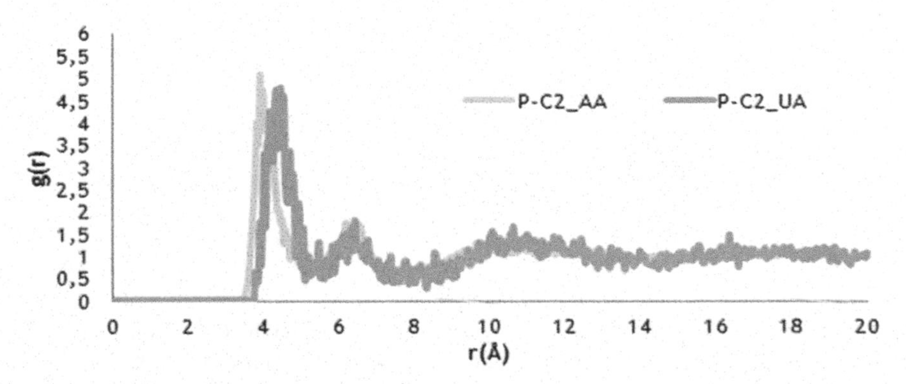

Figure 5.20 Radial distribution functions for the anion P atom and the C2 carbon atom of the imidazolium ring for the molecular description of AA and UA obtained at 298°K.

Then, the radial distribution functions (FDR) to the phosphorus atom (P) anion and the carbon atoms C2, C4, and C5 of the imidazolium ring is to 298°K.

Figures 5.20–5.22 present the FDR between the P atom in the anion and the carbon atoms C2, C4, and C5 in the imidazolium ring for molecular description of AA and UA. The two force fields used for molecular modeling of AA and UA show strong location atom (P) anion near the position C2, shown by the first large and sharp peak at approximately 4Å in the FDR. The first peak is at a slightly greater distance for the model for AA and AU, but with

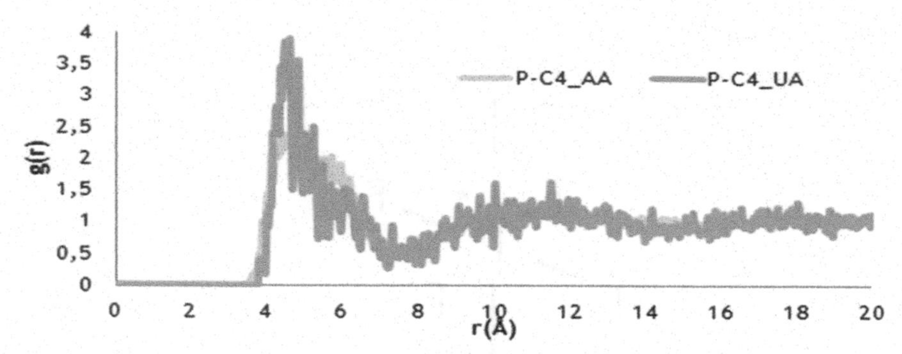

Figure 5.21 Radial distribution functions for the anion P atom and the C4 carbon atom of the imidazolium ring for the molecular description of AA and UA obtained at 298°K.

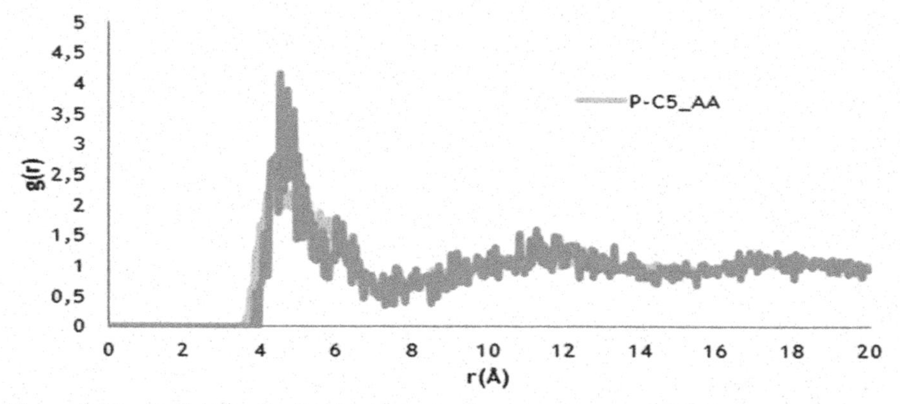

Figure 5.22 Radial distribution functions for the anion P atom and the C5 carbon atom of the imidazolium ring for the molecular description of AA and UA obtained at 298°K.

minor differences. As for the atom (P) of the anion with the other carbon of the imidazolium ring (C4 and C5), the differences between the peaks obtained by the two force fields are greater.

The force field of AA shows major difference in the first peak for the two carbon atoms, whereas UA shows sharper peaks. This indicates that the presence of the hydrogen atom attached to C4 and C5 carbons in AA force field exhibits anion association with the carbon atoms in the ring, reducing anion system on that side of the ring. Moreover, for the force field, UA locates a direct organization of the anion with the carbon atoms in the ring. After that peak, the FDR is similar for all force fields. The conclusions are given below.

5.4 Conclusions

The thermodynamic properties of the ionic liquid by molecular description of AA were obtained with deviations below the experimental data for the density 4.09. As for the determination of these properties using the molecular model UA values obtained, they showed deviations of 6.88%. For other deviations, thermodynamic properties on the experimental values obtained were very high. For determining the solubility of CO_2 in the IL, it was performed under two methods: explicit and Widom; however, Widom's method showed better results in the molecular description of AA.

As for the effect it had, the level of the molecular description of the potential for both AA and UA in determining thermodynamic properties of the system was tested as it was proved that the AA model has better results for the density, Henry constant, and isothermal compressibility constant compared to UA, confirming that the molecular model used AA or UA has an effect on the thermodynamic properties obtained. In conclusion, the determination of the thermodynamic properties of a system is possible to obtain very small deviation margins compared with experimental values, making prediction methods and computer simulation (MS) a fantastic tool for estimating properties *design materials* as *ionic liquids* and thus avoiding the great difficulty of synthesis in laboratory and cost for obtaining derivative.

Acknowledgments

To God for allowing me to live this. To my father (RIP) that although he is no longer with me, guide my way with his words of support that still live in my mind. To my family that always shown me support and unconditional love in everything I do. To you, thank you for the opportunity to realize this project which has now culminated in this publication and all those who were part of my academic education because thanks to you, I am the person and professional of today. I thank you with all my heart.

References

[1] Domínguez, E. (2004). Estudio del equilibrio trifásico mediante simulación molecular Montecarlo aplicado a mezclas binarias de agua con n-alcanos. Tesis Doctoral. Instituto Tecnológico de Celaya, México.

[2] Contreras, R. (2002). Determinación del Equilibrio Líquido-Vapor de Agua, Aromáticos y sus Mezclas mediante Simulación Molecular. Tesis Doctoral. Universitat Rovira I Virgili, España.

[3] Ungerer, P. B., Tavitian B., and Boutin, A. (2005). Applications of Molecular Simulation in the Oil and Gas Industry: Monte Carlo Methods. Editions Technip, Paris, France.

[4] Shah, J. K., and Maginn, E. J. (2004). Monte Carlo simulation study of the ionic liquid 1-n-Butyl-3-methylimidazolium hexafluorophosphate: liquid structure, volumetric properties and infinite dilution solution thermodynamics of CO_2. *Fluid Phase Equilibria*, 222, 195–203.

[5] Morrow, T., and Maginn, E. J. (2002). Molecular dynamics study of the ionic liquid 1-n-Butyl-3-methylimidazolium hexafluorophosphate. *J. Phys. Chem.* 106, 49, 12807–12813.

[6] MCCCS Towhee programa. Versión 7.0.1. http://towhee.sourceforge.net Agosto 2011

[7] Software Packmol. http://www.ime.unicamp.br/~martinez/packmol Agosto 2011

[8] Ghobadi, A. F., Taghikhani, V. and Elliott, J. R. (2011). Investigation on the solubility of SO_2 and CO_2 in imidazolium-based ionic liquids using NPT Monte Carlo Simulation. *J. Phys. Chem. B* 115, 13599–13607.

[9] Carvalho, P. J., and Coutinho, J. A. (2010). On the nonideality of CO_2 solutions in ionic liquids and other low volatile solvents. *J. Phys. Chem. Lett.* 1, 774–780.

[10] Gutiérrez, G. (2001). Elementos de simulación computacional. Manual de Dinámica molecular y Método Monte Carlo. Universidad de Santiago, Chile.

[11] Castellan, W. G. (1998). Fisicoquímica. Editorial Pearson, 2ª Edición.

[12] Colina, C. M., Olivera, F., Siperstein, F. R., Lisal, M. and Gubbins, K. E. (2003). Thermal properties of supercritical carbon dioxide by Monte Carlo simulations. *Mol. Simulat.* 29, 405–412.

[13] Engel, T., and Reid, P. (2007). Introducción a la Fisicoquímica: Termodinámica. Editorial Pearson, 1ª Edición.

[14] Zhao, W., Leroy, F., Heggen, B., Zahn, S., Kirchner, B., Balasubramanian S. and Müller-Plathe, F. (2009). Are There Stable Ion-Pairs in Room-Temperature Ionic Liquids? Molecular Dynamics Simulations of 1-n-Butyl-3-methylimidazolium Hexafluorophosphate. *J. Am. Chem. Soc.* 131, 15825–15833.

[15] Leach, R. Andrew. (2001). Molecular Modelling: Principles and Applications. Editorial Pearson, 2ª Edición.

[16] Potoff, J., and Siepmann, J. I. (2001). Vapor-liquid equilibria of mixtures containing alkanes, carbon dioxide, and nitrogen. *AIChE J.* 47, 1676–1682.

[17] Huang, X., Margulis, C. J., Li, Y., and Berne, B. J. (2005). Why is the partial molar volume of CO_2 so small when dissolved in a room temperature ionic liquid? Structure and dynamics of CO_2 dissolved in [Bmim+] [PF6-]. *J. Am. Chem. Soc.* 127, 17842–17851.

[18] Shah, J. K., Brennecke, F., and Maginn, E. J. (2002). Thermodynamic properties of the ionic liquid 1-n-Butyl-3-methylimidazolium hexafluorophosphate from Monte Carlo Simulations. *Green Chem.* 4, 112–118.

Index

About the Editor

Norma-Aurea Rangel-Vàzquez is a Professor Research at Technological Institute of Aguascalientes (ITA) in the Department of Metalmechanical in Materials Engineering program. She realized in 2007 the research of Polyurethane Porous/Hydroxyapatite for Biomedical Applications: Synthesis and Characterization for the degree of PhD in Polymers Science (Mexico). Later, she carried out a postdoctoral study using computational modeling at ITA. To date, She has have published several papers (30) in journals indexed in the Web of Science – Journal Citation Reports and in other journals, 1-book and 15-chapters book (Wiley, CRC-Press Taylor, ELSEVIER and, Science Publishing), 50 memories of national and international congress. She has served as a reviewer for different journals such as: Journal of Applied Polymer Science, Materials Letters, Carbohydrate Polymers, Journal Polymer Engineering & Science, Thermochimica Acta, Journal of Molecular Structure, Journal of Surface and Coatings Technology, Academic Journal of Agricultural Research, World Journal of Engineering and Physical Sciences (WJEPS), Journal of Agriculture Research, Annual Research & Review in Biology, Journal of Applied Physical Science International, Global Conference on Polymer and Composite Materials 2015 and International Journal of Nanomedicine. She had been participate in international committee in congress at: The 3rd Int'l Conference on Surface and Interface of Materials (SIM 2017), China. 2nd Conference International of Surface and Interface on Materials, Thailand. 2016 Spring World Congress on Engineering and Technologies (SCET 2016) in the topic: 2016 Spring International Conference on Materials Sciences and Technologies (MST-S), China. Global Conference on Polymer and Composite Materials 2015. China. She is on the Editorial Board Member of several journals including, Journal of Pharmaceutics & Drug Development, Material Science and Engineering with Advanced Research, Journal of Nanomedicine Research, Nanomedicine Research Journal, BAOJ Pharmaceutical Science (Bioaccent) and Chemistry Journal. Currently. Currently, she is Editor in Chief in Journal of Pharmaceutics & Drug Development and BAOJ Pharmaceutical Science (Bioaccent). She is interested on Controlled drug delivery systems. In particular, skin, diabetes mellitus, antifungical, Design and preparation of oral solid (tablets and capsules) dosage forms, Analysis of drugs in pure forms, formulations and biological matrices, Bioadhesive drug delivery systems, especially hydrogels Polymers, biopolymers and IPNs, respectively.

Lightning Source UK Ltd.
Milton Keynes UK
UKOW06n1146270117
293027UK00002B/9/P

9 788793 379954